Programmable Logic Controllers and their Engineering Applications

D0319269

Alan J Crispin

BICC Industrial Reader
Division of Engineering Faculty of Information and Engineering Systems
Leeds Polytechnic

07

McGRAW-HILL BOOK COMPANY

London · New York · St Louis · San Francisco · Auckland · Bogotá
Guatemala · Hamburg · Lisbon · Madrid · Mexico · Montreal · New Delhi
Panama · Paris · San Juan · São Paulo · Singapore · Sydney · Tokyo · Toronto

Published by

McGRAW-HILL Book Company (UK) Limited

SHOPPENHANGERS ROAD · MAIDENHEAD · BERKSHIRE · ENGLAND

TEL: 0628-23432; FAX: 0628-35895

British Library Cataloguing in Publication Data
Crispin, Alan J. (Alan John), *1959–*
    Programmable logic controllers and their engineering
    applications.
    1. Programmable automatic control systems
    I. Title
    629.8′95

    ISBN 0–07–707227–8

Library of Congress Cataloging-in-Publication Data
Crispin, Alan J., 1959–
    Programmable logic controllers and their engineering
    applications
    Alan J. Crispin.
       p. cm.
    ISBN 0–07–707227–8
    1. Programmable controllers. I. Title.
    TJ223.P76C75   1990
    629.8′95–dc20   89–13949
                                                                        CIP

1234MAT9210

Typeset by TecSet Limited, Wallington, Surrey and printed and bound by M.A. Thomson Litho Ltd, Glasgow

Contents

Preface

A modern programmable logic controller (PLC) is a simple control system which is easy to use and versatile. Most automated factories employ programmable logic controllers in the control of production and assembly processes.

The first commercially successful PLC was developed in 1969 by Modicon (now part of Gould Electronics) for General Motors' Hydramatic Division. Since then, companies such as Allen Bradley, General Electric, GEC, Siemens and Westinghouse have developed a range of medium-cost, high-performance PLCs. The low-cost PLC market is dominated by Japanese companies such as Mitsubishi, Omron and Toshiba.

The aim of this book is to provide an introduction to the design, structure and operation of PLCs and an insight into their applications. The book has been written for both the practising engineer and the student studying automation and control. It contains many worked examples and case studies which should prove invaluable to both.

Chapter 1 describes the basics of PLCs and their logic-based operation. It introduces many of the topics that are covered in detail in later chapters.

Chapters 2–4 deal with hardware, i.e. physical components. The operation of the hardware internal to programmable systems is discussed in Chapter 2. Chapter 3 shows how to interface digital sensors and actuators whereas Chapter 4 deals with analogue devices and includes a discussion on data converters.

Chapters 5 and 6 deal with ladder programming which is the programming technique used to program most PLCs. Numerous programming examples are given.

Applications are covered in Chapters 7 and 8. The examples of Chapter 7 can be set up in the laboratory. The case studies of Chapter 8 discuss typical industrial applications. Ladder programs are developed for the industrial case studies.

The final chapter deals with factory communications. Standard serial and parallel data exchange methods together with local area networks are discussed.

I would like to extend my thanks to all the people who have in any way contributed to this book, in particular, Bob Ward, Pam Butterworth and Don Crispin.

1
Introduction

1.1 Basics of programmable logic controllers

An apt definition of a programmable logic controller (PLC) is that it is a 'digital electronic device that uses a programmable memory to store instructions and to implement specific functions such as logic, sequence, timing, counting and arithmetic to control machines and processes'.[1]

Figure 1.1 shows how the control action is achieved. Input devices (e.g. mechanical contacts, proximity switches) and output devices (e.g. motors, solenoids) from the machine or process to be controlled are connected to the PLC. A user enters a sequence of instructions (known as the program) into the PLC's program memory. The controller then

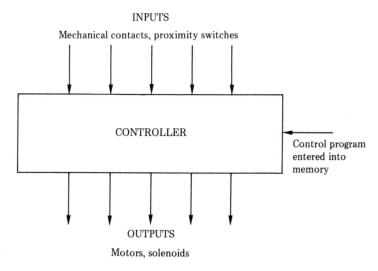

Figure 1.1 The control action of a PLC.

[1] National Electrical Manufacturers' Association (USA).

1

continuously monitors the state of the inputs and switches outputs according to the user's program.

Because the stored program can be modified, new control features can be added or old ones changed without rewiring the input and output devices. The result is a flexible system which can be used for control tasks that vary in nature and complexity.

The main differences between a PLC and, say, a microcomputer are that:

1. Programming is predominantly concerned with logic and switching operations.
2. The interfacing for input and output devices is inside the controller.
3. They are rugged being packaged to withstand vibration, temperature, humidity and noise.

Some typical programmable logic controllers are shown in Figure 1.2. The main body of a PLC is called the base unit and it contains a central processing unit, the program memory and the necessary interfacing for input/output devices. The central processing unit controls the overall operation of the system and is discussed further in Chapter 2. A separate component called a program loader or console is used to enter instructions into the program memory.

The term 'logic' features in the name 'programmable logic controller' because programming is based on the logic demands of input devices. Programs implemented are predominantly logical rather than discrete versions of continuous algorithms. The following section provides an introduction to logic.

1.2 Logic

In the field of logic we are normally concerned with systems that work on a straightforward two-state basis. The two states might be the ON and OFF positions of a single switch. Alternatively, they might be the open and shut positions of a pneumatic valve.

We use Boolean algebra when dealing with logic systems. This is a form of logical algebra named after the nineteenth-century mathematician Rev. George Boole who first developed the subject. In Boolean terms, the two states of a single switch correspond to the truth values TRUE and FALSE or 1 and 0. If we label a switch with letter A, it is then possible to write $A = 1$ when the switch is ON and $A = 0$ when the switch is OFF. The letter A is called a Boolean variable.

Single-pole, single-throw switches such as those shown in Fig. 1.3(a) are of two types: those with contacts which are normally open (abbreviated to NO) and those with contacts which are normally closed (abbreviated to NC). If we use a Boolean variable A to represent a normally open switch we denote a normally closed switch by NOT A or

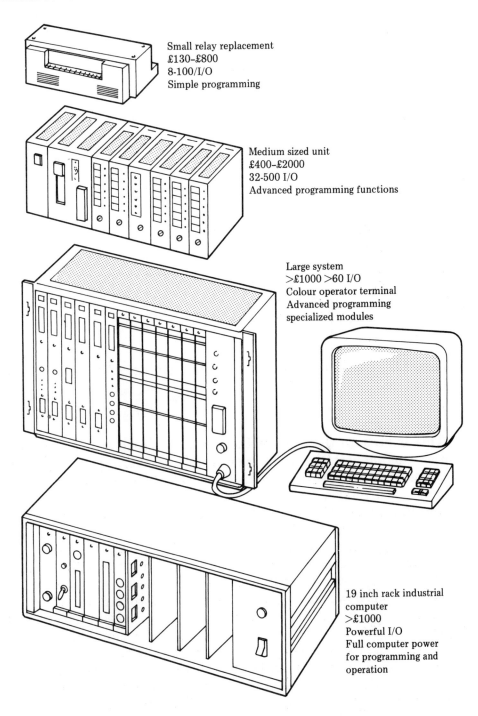

Small relay replacement
£130–£800
8-100/I/O
Simple programming

Medium sized unit
£400–£2000
32-500 I/O
Advanced programming functions

Large system
>£1000 >60 I/O
Colour operator terminal
Advanced programming
specialized modules

19 inch rack industrial
computer
>£1000
Powerful I/O
Full computer power
for programming and
operation

Figure 1.2 Types of PLC with approximate cost indicated. *Source*:
Florence, D. H., 'PLCs aid low cost automation', *Professional
Engineering*, July 1988, p.32.

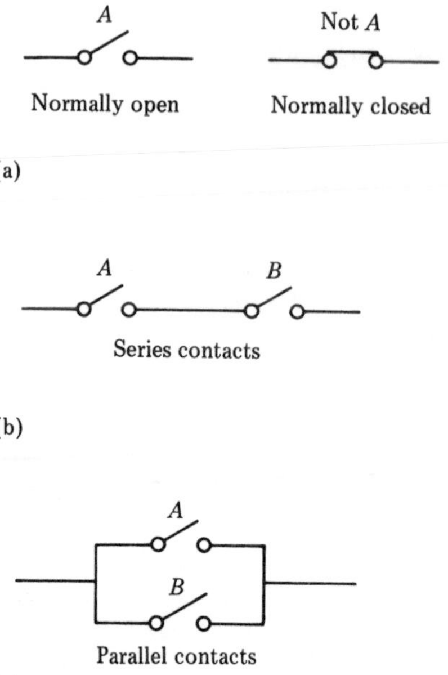

(a)

A	Not A
0	1
1	0

(b)

A	B	A.B
0	0	0
0	1	0
1	0	0
1	1	1

(c)

A	B	A + B
0	0	0
0	1	1
1	0	1
1	1	1

Figure 1.3   Logic functions; (a) NOT function; (b) AND function; (c) OR function.

$\bar{A}$. This is because the state of the normally closed switch is always opposite to that of the normally open switch. An overbar is always used to denote inversion so that $\bar{B}$ means NOT B, NOT being the logic function for inversion.

The truth values (i.e. 0 and 1) of NOT A are always the inverse of A so we define NOT A by tabulating its truth values against those of A as shown in Fig. 1.3. A table of this kind is called a truth table.

The AND logic function describes the operation of two normally open single-pole, single-throw switches connected in series as shown in Fig. 1.3(b). Current flows only when switch A AND switch B are closed, i.e. when $A = 1$ and $B = 1$. In Boolean algebra, a dot is used to mean AND. Thus, $A.B$ means A AND B. The truth table for the AND function establishes the relationship between A, B and the result $A.B$.

The OR logic function describes the operation of two normally open switches connected in parallel as shown in Fig. 1.3(c). Current flows when switch A OR switch B, or both, is closed. The Boolean notation for A OR B is $A + B$. The truth table for the OR function establishes the relationship between A, B and the result $A + B$.

The laws of Boolean algebra allow us to find alternative forms for a switch circuit. We shall find that some of these alternative forms are either easier to program or can be used to eliminate redundant switches. The laws, introduced in Fig. 1.4 below, are in the form of logic equations and can be verified with the switching arrangements drawn at the sides.

In terms of switch contacts, the commutative laws, labelled (a), emphasize the fact that the order in which two open switches are connected in parallel (or in series) has no effect on the overall result. The associative laws (b) show the equivalence of switch groupings involving three open switches. The switch representations of the distributive laws (c) show that equivalence can be obtained by introduc-

$$A + B = B + A$$

$$A \cdot B = B \cdot A$$

(a) Commutative laws

$$A + (B + C) = (A + B) + C$$

$$A \cdot (B \cdot C) = (A \cdot B) \cdot C$$

(b) Associative laws

$$A \cdot (B + C) = A \cdot B + B \cdot C$$

$$A + (B \cdot C) = (A + B) \cdot (A + C)$$

(c) Distributive laws

$$A + (A + B) = A$$

$$A + (A \cdot B) = A$$

(d) Absorption laws

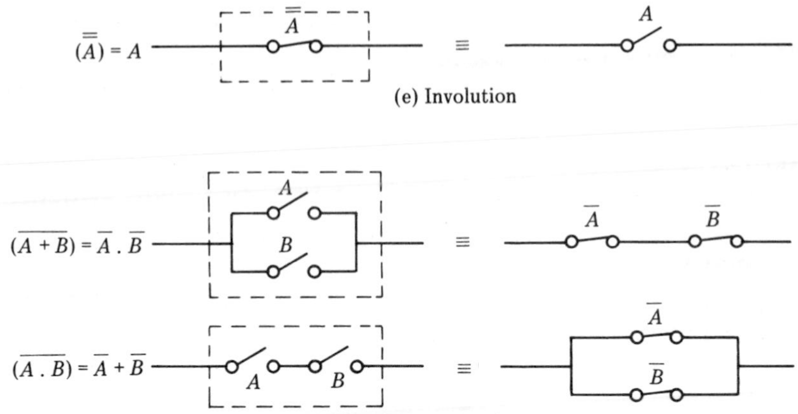

$$(\overline{\overline{A}}) = A$$

(e) Involution

$$(\overline{A + B}) = \overline{A} \cdot \overline{B}$$

$$(\overline{A \cdot B}) = \overline{A} + \overline{B}$$

(f) Inversion laws
(De Morgan's theorems)

Figure 1.4   Laws of Boolean algebra; (a) commutative; (b) associative; (c) distributive; (d) absorption; (e) involution; (f) inversion (De Morgan's theorems).

ing redundancy. The logic for the switches in the right arrangements is identical to the left arrangements but the right arrangements use an extra (redundant) contact for switch A. A pair of contacts labelled with the same Boolean variable, for example A, are ganged so that they operate in unison.

In terms of switch arrangements, the absorption laws (d) minimize circuits which use a ganged switch. Because switch A is ganged, switch B is redundant. The switch representation of the involution law (e) shows that the inverse of a normally closed switch is a normally open switch. The inversion laws (also called De Morgan's theorems), labelled (f), are used when it is required to invert the Boolean functions $A + B$ (A OR B) and $A.B$ (A AND B). We call $\overline{A.B}$ the NAND logic function (the name NAND is derived from NOT AND) $\overline{A+B}$ the NOR logic function (the name NOR is derived from NOT OR).

The theory discussed so far has been related to switch contacts but applies to any logic system constructed from logic elements. In digital electronics, for example, integrated circuits form the logic elements which are called gates. These are represented by the functional logic symbols shown in Fig. 1.5.

1.3   PLC layout

As we have seen, a switch is a logic element and so can be used to provide a logic signal input to a PLC. This section discusses how simple

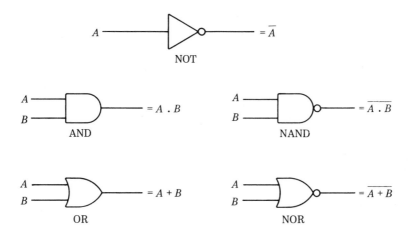

Figure 1.5 Logic gate symbols.

input devices such as switches, and output devices are connected to a PLC base unit.

A block diagram showing a typical base unit arrangement is shown in Fig. 1.6. Each PLC input is energized (turned on) when 24 V d.c. is applied to it from a switching device. Normally a 24 V d.c. supply is internally generated from the mains input and is used for wiring up input devices. Switches that are connected to the input lines can be of

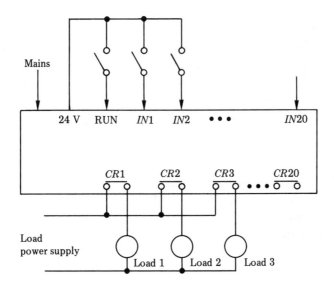

Figure 1.6 Base unit arrangement.

the normally open or normally closed contact type. When the run input is energized, the outputs are switched according to the program and the condition of the inputs.

The output loads can be switched from relay, transistor or triac contacts inside the PLC. Relays are widely used. Figure 1.6 shows how loads such as solenoids, motors and heaters can be connected to relay contacts. This is fine provided that the maximum current rating for the relay contacts is not exceeded. For a heavy current load, a PLC output relay is used to drive a secondary switching device such as a solid-state relay or a contactor. Secondary switching devices are discussed further in Chapter 3.

The input and output connection points on a PLC are called ports. The ports are allocated numbers so that they can be uniquely identified. Each PLC manufacturer uses his own identification system which depends on the number of input/output options. In Fig. 1.6 the input ports, referred to as contacts, are labelled *IN* and numbered 1 to 20. The output ports are labelled with a *CR* (*CR* is an abbreviation for control relay) and numbered 1 to 20. Other types of port are used for handling analogue signals, high-speed pulses and communications.

1.4 Ladder logic

With the majority of programmable logic controllers, writing a program is equivalent to drawing a switching circuit. The switching circuit is drawn in a ladder diagram format. This format requires that:

1. Circuits are arranged as a series of horizontal lines containing inputs (referred to as contacts) and outputs (referred to as coils). Typical circuit lines are shown in Fig. 1.7.
2. Inputs must always precede outputs and are in the form of normally open and normally closed contacts. The ladder symbol for a normally open contact is | |. The symbol for a normally closed contact is |/|.
3. There must always be at least one output on each line. An output is, for example, a PLC output relay. The ladder symbol for a PLC output is drawn either as two parentheses close together, i.e. ( ), or as a circle.
4. Circuits in the form of vertical lines are not used.
5. The numerical assignments for the inputs (contacts) and outputs (coils) form part of the ladder diagram.
6. Other elements such as timers, counters and shift registers can be implemented in ladder diagrams. These elements are discussed in Chapter 5.

The term ladder is used because the lines of a completed diagram resemble the rungs of a ladder (see Fig. 1.7). The two vertical lines are called bus lines and represent the power connections, in this case 24 V

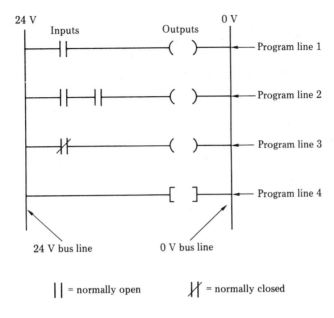

Figure 1.7 Ladder format.

and 0 V. Each horizontal line represents a program line. The output on a program line is energized (turned on) when the input contact(s) to it are made (i.e. when the contacts connect the 24 V supply to the coil).

A ladder diagram can be translated into a program consisting of instructions and data. Table 1.1 describes some Boolean instructions that are used by PLC manufacturers. Ladder programs are entered into memory in an address–instruction–data format. An address is a number which activates a memory location. Instructions and data are entered into sequential memory locations usually starting from address zero.

Table 1.1 Ladder instructions

Instruction	Description
LOAD	Load logical state of start input
LOAD NOT	Load logical state of start input and invert
AND	Logical AND operation
AND NOT	Logical AND NOT operation
OR	Logical OR operation
OR NOT	Logical OR NOT operation
OUT	Output

It is always the ladder logic which determines how the outputs are energized. Consider Fig. 1.8, which shows two controllers both having a normally open switch connected to their input ports $IN1$. The controller shown in Fig. 1.8(a) is ladder programmed so that when the switch connected to $IN1$ closes, the output load connected to $CR1$ is energized (turned on). The controller shown in Fig. 1.8(b) is ladder programmed so that when the switch connected to $IN1$ closes the output load connected to $CR1$ is de-energized (turned off). The control action of Fig. 1.8(b) is opposite to that of Fig. 1.8(a) because the NOT function is used in the ladder. In Fig. 1.8(a) we load the logical state of the input and use this to control the output. In Fig. 1.8(b) we read the logical state of the input and invert its value, which we then use to control the output.

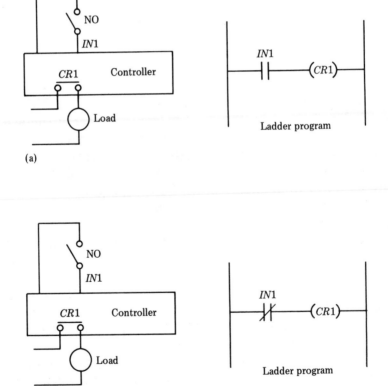

(a)

(b)

Figure 1.8 Ladder control action.

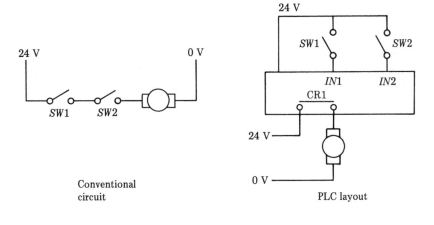

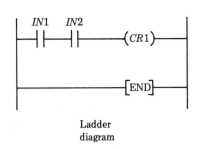

Ladder
diagram

Address	Instruction	Data
0	LOAD	$IN1$
1	AND	$IN2$
2	OUT	$CR1$
3	END	

Program

Figure 1.9 PLC implementation of a series circuit.

An example of how a PLC may be used to implement a conventional control circuit is shown in Fig. 1.9. The conventional circuit uses two normally open switches in series to energize a motor. If this action is to be implemented using a PLC, the switches are connected to input ports, in this case $IN1$ and $IN2$, and the motor to an output relay such as $CR1$. The control action is then described by the ladder diagram and program shown in Fig. 1.9 which is based on an AND function.

In Fig. 1.10, a motor is controlled by using two normally open switches in parallel. If this action is to be implemented using a PLC, the switches are connected to input ports, in this case $IN1$ and $IN2$, and the motor to an output relay such as $CR1$. The control action is then described by the ladder diagram and program shown in Fig. 1.10, which is based on an OR function.

When drawing ladder diagrams, you are allowed to use internal feedback, i.e. use output relay contacts as inputs. Inputs and outputs may represent the state of timers, counters and internal coils and flags instead of input/output ports. Ladder programming is discussed further in Chapter 5.

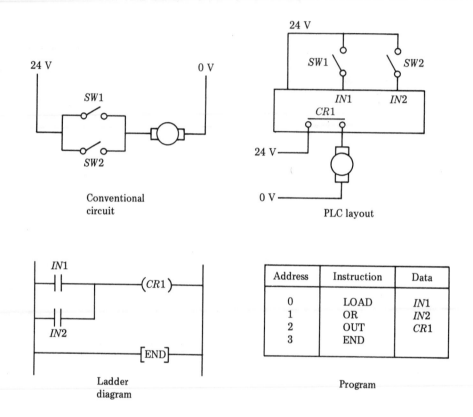

Figure 1.10 PLC implementation of a parallel circuit.

Questions

1. Which logic function is used for inversion?
2. Which logic function is used for switches connected in series?
3. Which logic function is used for switches connected in parallel?
4. Draw the ladder symbols which represent normally open and normally closed contacts.
5. The switching circuit shown in Fig. 1.Q5 is to be implemented using a PLC. Show how the controller's inputs and output can be arranged. Write the ladder program.

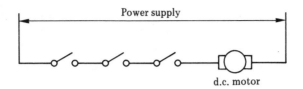

Figure 1.Q5 Multiple AND operation.

6. The PLC shown in Fig. 1.Q6 is to be used as a controller which implements NAND control action. The truth table for the NAND logic function is also shown. Using De Morgan's theorem or other method, develop a simple ladder diagram which achieves this.

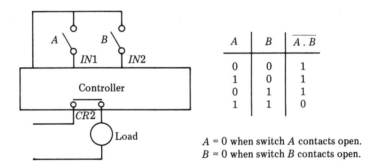

A	B	$\overline{A \cdot B}$
0	0	1
1	0	1
0	1	1
1	1	0

$A = 0$ when switch A contacts open.
$B = 0$ when switch B contacts open.

Figure 1.Q6 NAND.

7. The PLC shown in Fig. 1.Q7 is to be used as a controller which implements a NOR control action. The truth table for the NOR logic function is also shown. Using De Morgan's theorem or other method develop a simple ladder diagram which achieves this.

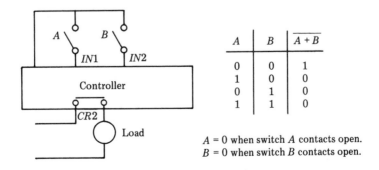

A	B	$\overline{A + B}$
0	0	1
1	0	0
0	1	0
1	1	0

$A = 0$ when switch A contacts open.
$B = 0$ when switch B contacts open.

Figure 1.Q7 NOR.

8. The exclusive OR function (written as XOR) is a special form of the OR function. Its truth table is as shown in Fig. 1.Q8. When the inputs to an XOR device are alike the output is logic O and when the inputs are unalike the output is logic 1. Show how a PLC can be used as a controller that implements XOR action.

A	B	$A \oplus B$
0	0	0
1	0	1
0	1	1
1	1	0

$\oplus$ = Logic symbol for XOR function

Figure 1.Q8   XOR

2

Design, structure and operation

2.1 Introduction

In this chapter we focus our attention on the design, structure and operation of a programmable logic controller. All PLCs have a CPU, memory and interface circuits. The internal structure of a PLC is called its architecture.

2.2 Basic structure and operation of a PLC

A block diagram of the internal structure of a PLC is shown in Fig. 2.1. The blocks consist of a central processing unit (CPU), a main memory and a buffer consisting of image memory and connection circuitry for digital input/output devices. A communications bus (i.e. a group of parallel wires used for transmitting digital signals) forms a common link to allow each element to share information.

The input image memory is used to hold the on/off states of individual input ports. We use the binary system to represent these on/off states because it is based on two digits (1 and O). In image memory, an on state is stored as a binary 1 and an off state is stored as a binary O.

The CPU processes the binary data stored in the input image memory and corresponding data held in the output image memory according to the user's program which is stored in the main memory. The bit values held in the output image memory determine which output ports are energized. A binary 1 sets an output port on and a binary O sets an output port off.

A special program called the operating system controls the actions of the CPU and consequently the execution of the user's program. The operating system is supplied by the PLC manufacturer and is permanently held in memory. A PLC operating system is designed to scan image memory and the main memory which stores the ladder diagram program.

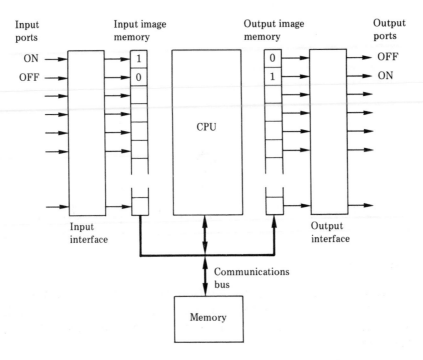

Figure 2.1   Internal structure of a PLC.

It should be clear from this preamble that we use the memory to store various types of information. This information might be an image of the input/output ports, the user's program, the operating system or data. Different types of memory device are used for different types of information.

2.3 Memory

Memory is characterized by its volatility. A memory is volatile if it loses its data when the power to it is switched off and non-volatile otherwise. Common types of memory include semiconductor memory and magnetic disk. The various types of semiconductor memory are:

1. *RAM* Random access memory is a flexible type of read/write memory. All PLCs will have some amount of RAM, which is used to store ladder programs being developed by the user, program data which needs to be modified and image data.

   RAM is volatile. This means that RAM cannot be used to store data while the PLC is turned off unless the RAM is battery backed. A type of RAM called CMOS RAM (complementary metal-oxide semiconductor RAM) is suitable for use with batteries because it consumes very little power and operates over a very wide range of supply voltages.

2. *ROM*   A read only memory is programmed during its manufacture using a mask. It is a non-volatile memory and provides permanent storage for the operating system and fixed data.

3. *EPROM*   Erasable programmable read only memory is a type of ROM which can be programmed by electrical pulses and erased by exposing a transparent quartz window found in the top of each device to ultraviolet light. EPROM is non-volatile memory and provides permanent storage for ladder programs.

4. *EEPROM*   Electrically erasable programmable read only memory is similar to EPROM but is erased by using electrical pulses rather than ultraviolet light. It has the flexibility of battery-backed CMOS RAM. However, writing data into an EEPROM takes much longer than into a RAM.

2.3.1   MEMORY STORAGE CAPACITY

The storage capacity of a memory device is determined by the number of binary digits, i.e. on/off states, it can hold. In microelectronics, 1K represents the number 1024 i.e. the binary number 2^{10}. A 4K byte memory is capable of storing 4×1024 words, each of 8 bits, and has a total storage capacity of 32 768 bits.

Clearly, the storage capacity of the user memory will determine the maximum program size. As a guide, a 1K byte memory will hold 1024 program instructions and data if these are stored as groups of 8 bits.

2.3.2   MEMORY MAP

We use the term memory mapping to describe the situation in which input/output ports are controlled by writing data into the image memory. A diagram which shows the allocation of memory addresses of ROM, RAM and I/O is called a memory map. Figure 2.2 illustrates a memory map for a typical PLC. In this, image bits are stored in RAM above the user's program and data for flags, counters and timers. Flags, counters and timers are discussed below. With most PLCs the memory map is already configured by the manufacturer. This means that the program capacity, the number of input/output ports and the number of internal flags, counters and timers are fixed.

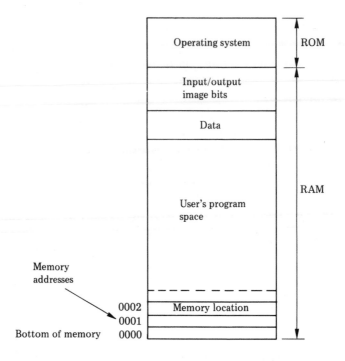

Figure 2.2   Memory map.

2.4   Program consoles

Programs are entered into the PLC's memory using a program console (loader). Program consoles vary from hand-held systems incorporating a small keyboard and liquid crystal displays (LCDs) to CRT (cathode ray tube) terminals. Typical hand-held PLC program consoles for entering ladder code and diagrams are shown in Fig. 2.3

2.5   Central processing unit (CPU)

Invariably, the central processing unit of a PLC will be built up around a microprocessor which is a data-processing circuit scaled down to fit on a single silicon chip. The function of the CPU is to accept data in the form of groups of binary digits and perform arithmetic and logical operations on the data in accordance with instructions stored in the memory.

The internal structure of a CPU comprises input and output interfaces, a memory in the form of registers and a control element called the arithmetic and logic unit (ALU). The input and output interfaces allow the CPU to read data from the memory and write data to the memory via the communications bus. The ALU performs arithmetical and logical operations on data stored in the CPU registers.

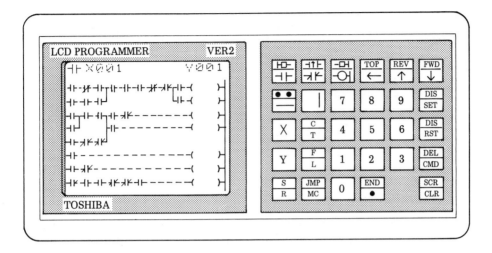

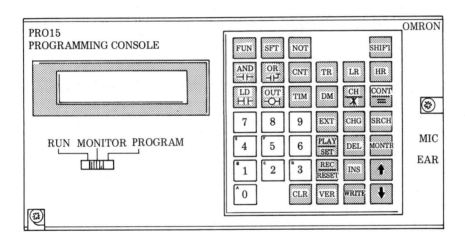

Figure 2.3 Typical PLC program consoles.

2.5.1 REGISTERS

Most CPU operations involve the use of a register, which is a memory element used to store a group of bits on a temporary basis. CPU registers are located inside the microprocessor. So-called data registers are located in RAM and are used for storing flags, counter and timer constants and other types of data. Figure 2.4 illustrates examples of registers having different storage capacities. A 4-bit register stores a *nibble*, which is 4 bits of data. An 8-bit register stores a *byte*, which is 8 bits of data. A 16-bit register stores a *word*, which is 16 bits of data.

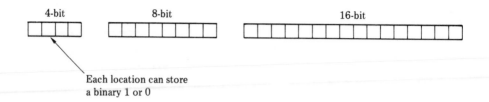

Figure 2.4 Registers.

2.5.2 FLAG REGISTERS

If a bit state (0 or 1) is used to indicate that some condition has occurred it is called a flag. A register which stores a group of flag bits is called a flag register. The CPU has an internal flag register which contains information about the result of the latest arithmetical and logical operations. PLC image memory is effectively a flag register, as it contains the current status of the inputs and outputs.

2.5.3 AUXILIARY RELAYS

Auxiliary relays are single-bit memory elements located in RAM that may be manipulated by the user's program. They are called auxiliary relays because they may be likened to imaginary internal relays. A battery-backed auxiliary relay is called a retentive or holding relay and can be used for storing data during power failure. A number of auxiliary relays may be grouped together to form a register.

It is important to remember that because auxiliary relays are only bit values stored in memory output loads cannot be connected directly to them. However, auxiliary relays can be used to control output loads indirectly.

2.5.4 SHIFT REGISTERS

Some registers are arranged so that bits stored in them can be moved one position to the left or to the right with the application of a shift command or pulse. Such registers are called shift registers and can be used for sequence control applications (see Chapter 5).

Figure 2.5 illustrates the action of a shift register. An on state (binary 1) is fed into an 8-bit register and then shifted stage by stage through the register to the other end by the application of a series of shift-right signals. A further shift-right command would push the binary 1 off the end of the register where it would normally be lost. Data bits could have been fed into the shift register at any stage and so we are not limited to just advancing a binary 1 through eight register positions.

Data input	Shift register	Shift signals
1	1 0 0 0 0 0 0 0	1st
0	0 1 0 0 0 0 0 0	2nd
0	0 0 1 0 0 0 0 0	3rd
0	0 0 0 1 0 0 0 0	4th
0	0 0 0 0 1 0 0 0	5th
0	0 0 0 0 0 1 0 0	6th
0	0 0 0 0 0 0 1 0	7th
0	0 0 0 0 0 0 0 1	8th

Figure 2.5   Shift register.

2.5.5 BINARY COUNTER

The CPU may function as a binary counter since it is able to increment (add one to) and decrement (subtract one from) binary data stored in a register and compare binary data stored in two separate registers. Counters are used to count, for example, digital pulses generated from a switching device connected to an input port. An output is usually generated after a predetermined number of input pulses have been counted. The count value required is stored in a data register.

2.5.6 TIMERS

A CPU will have a built-in clock oscillator which controls the rate at which it operates. The CPU uses the clock signal to generate delay times. A delay time could be used, for example, to keep an output relay energized for a fixed period.

2.6 Number systems

The CPU processes binary data stored in registers. Since writing down groups of binary digits is rather cumbersome we either convert them to decimal numbers or use other number systems such as octal, hexadecimal and binary-coded decimal (BCD).

2.6.1 BINARY

The binary system uses two counting digits (1 and 0). We write a binary number with its most significant bit (MSB) furthest to the left and its least significant bit (LSB) furthest to the right so that it is organized in ascending powers of 2 starting with the LSB. Converting a binary number into an equivalent decimal number is quite straightforward, as illustrated in Fig. 2.6.

Bit 7 (MSB)   Bit 0 (LSB)

1	0	1	1	0	1	0	1

128 64   32   16   8   4   2   U
2^7 2^6 2^5 2^4 2^3 2^2 2^1 2^0

$10110101 = (1 \times 128) + (0 \times 64) + (1 \times 32) + (1 \times 16) + (0 \times 8) + (1 \times 4) + (0 \times 2) + (1 \times 1)$
$= 181$ decimal.

Figure 2.6   Binary conversion.

2.6.2   OCTAL

The octal (base 8) number system uses a set of eight distinct digits, 0 to 7. Octal numbers are organized in ascending powers of 8. For example, the octal number $321 = 3 \times 8^2 + 2 \times 8^1 + 1 \times 8^0 = 209$ decimal.

Converting a binary number to octal requires that it is split up into 3-bit groups. This is because the octal numbers 0 to 7 represent 3-bit binary numbers ranging from 000 to 111. Figure 2.7 illustrates how a binary number is converted to octal and decimal.

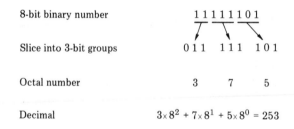

8-bit binary number   1 1 1 1 1 1 0 1

Slice into 3-bit groups   0 1 1   1 1 1   1 0 1

Octal number   3   7   5

Decimal $3 \times 8^2 + 7 \times 8^1 + 5 \times 8^0 = 253$

Figure 2.7   Binary to octal to decimal.

2.6.3   HEXADECIMAL

The hexadecimal number system (base 16) uses a set of sixteen digits (0, 1, 2, 3, 4, 5, 6, 7, 8, 9, A, B, C, D, E, F). The letters A to F represent the numbers 10 to 15. Hexadecimal numbers are organized in ascending powers of 16. For example, the hexadecimal number $321 = 3 \times 16^2 + 2 \times 16^1 + 1 \times 16^0 = 801$ decimal.

Converting a binary number to hexadecimal requires that it is split up into 4-bit groups (i.e. nibbles). This is because the hexadecimal numbers 0 to F represent 4-bit binary numbers ranging from 0000 to 1111. Figure 2.8 shows how to convert a binary number to hexadecimal and decimal.

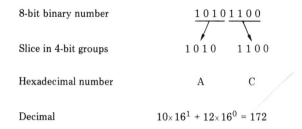

Figure 2.8 Binary to hexadecimal to decimal.

2.6.4 BINARY-CODED DECIMAL

The binary-coded decimal system (BCD) codes decimal numbers into 4-bit binary numbers. The decimal digits 0 to 9 are represented by the binary numbers 0000 to 1001. Figure 2.9 shows how to convert a decimal number to BCD.

Figure 2.9 BCD conversion.

2.7 PLC operating system

All PLC operating systems execute a ladder program by scanning the logic states of the inputs and outputs stored in image memory. Figure 2.10 shows how this is done. The operating system first scans all the inputs. This may be done either column by column, as shown, or rung by rung. The logic is then solved. Finally the outputs are scanned and switched according to the state of the inputs and the program logic. The run loop repeats the scanning action over and over again.

The program logic might involve simple AND, OR, NOT functions as shown in Fig. 2.10 or more advanced counting, timing, sequence and mathematical functions. The operating system determines what functions are available to the user. The more sophisticated the operating system the more programming functions are provided.

The operating system is characterized by:

1. *Scan rate* The speed at which a PLC scans the memory is called the scan rate. The scan rate depends on how fast the CPU is clocked. It is expressed in terms of how many seconds it takes to scan a given amount of memory, usually 1K bytes.

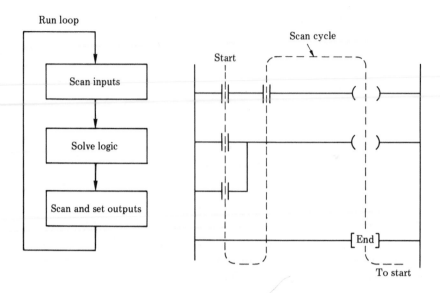

Figure 2.10   Executing a ladder program.

The actual time to scan a program will depend on the scan rate, the length of the program and the types of functions used in a program. The faster the scan time the more often the inputs and outputs are checked.

2. *Phasing errors*   The CPU, under the control of the operating system, scans the input image memory rather than the inputs themselves. The input memory is not changed while the CPU is scanning it. Thus it is possible for an input port to change its state from, say, off to on to off again before the input image memory is updated. A phasing error is said to have occurred when the CPU scan misses a change of state of an input port.

2.8   Multitasking

More advanced PLCs use multitasking. This is the process of running two or more control tasks using a single CPU. Each task has its own program and allocated input/output ports. The CPU may schedule its processing time among the various tasks (this is called time-driven multitasking) or allow events to initiate the various tasks (this is called event-driven multitasking). Tasks are assigned priority levels. Higher-priority tasks are always executed before lower-priority tasks.

Multitasking systems make use of interrupts. An interrupt is a special control signal to the CPU which tells it to stop executing the program in hand and start executing another program stored elsewhere in memory. The CPU clock oscillator can be used to provide interrupts at regular intervals so that processor time can be shared between tasks (time-driven multitasking). Alternatively, an external event such as a machine fault alarm can be used to drive the interrupt line (event-driven multitasking).

2.9 Types of Ports

2.9.1 OPTO-ISOLATED DIGITAL INPUTS

A base unit input interface circuit will use an opto-isolator arrangement such as that shown in Fig. 2.11. An opto-isolator is a device which uses light to couple signals from one system to another; in this case the input device and the image memory circuit. It incorporates a light-emitting diode (LED) and phototransistor for this purpose. The device provides a very large degree of electrical isolation betwen the two systems.

The circuit shown in fig. 2.11 incorporates a status LED and potential divider. The status LED tells the user the current logic state of an input line. Potential dividers are used to convert a high voltage, say 24 V, to a lower voltage, say 5 V. This is required because semiconductor memory

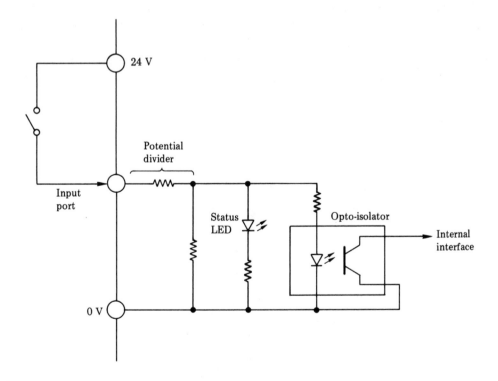

Figure 2.11 Opto-isolated input interface.

uses the voltage levels 5 V and 0 V rather than 24 V and 0 V for the bit states 1 and 0 respectively.

Input interfaces can be divided into those which require the input device to source current and those which require the input device to sink current. The interface of Fig. 2.10 requires that the input device sources current from the power supply. Other types of interface require that the input device sinks current to the ground terminal.

2.9.2   RELAY OUTPUTS

A traditional relay is a switch controlled by an electromagnet. Relays are used in PLCs because they can handle large currents and offer a high degree of isolation between the PLC circuits and the load circuits. A typical relay will be capable of switching a few amperes. However, relays have the following disadvantages: (a) they are slow to operate, (b) when closed their contacts can bounce before settling and (c) the relay coil can generate large inductive currents when it is energized.

A typical electromagnetic relay-based output circuit is shown in Fig. 2.12. The npn transistor switches current through the relay coil to close its contacts. The transistor is controlled by the image memory circuit. The diode is connected across the relay coil in order to protect the transistor from the effects of back e.m.f. This is the reverse voltage developed in the relay coil which causes the unwanted inductive current which opposes the normal flow of current.

On a practical note, it is important not to exceed the maximum current that the output relay contacts can handle. For d.c. loads the

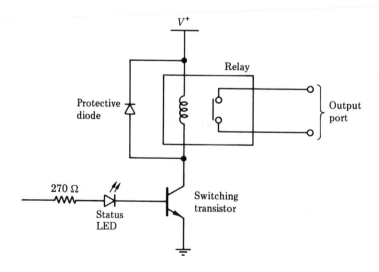

Figure 2.12   Relay output stage.

current rating will be given in amperes. For a.c. loads the maximum current will be given as a VA (volts × amperes) power rating.

For example, if the specification of the output relays is given as 35 VA, then the maximum current the contacts can handle at 240 V $_{r.m.s.}$ is calculated as

$$I = 35/240 = 125 \times 10^{-3} = 146 \text{ mA}$$

If the load current is likely to exceed the rated maximum current of the internal relay, then an external secondary heavy-duty relay should be used.

A solid-state relay is not an electromagnetic device but rather a device that uses an optically isolated triac. These have many advantages over a traditional relay and are discussed in Chapter 3.

2.9.3 TRANSISTOR OUTPUTS

A transistor can be operated as an electrically controlled switch. In this condition, a sufficient current applied to its base enables a saturated current to flow through the collector–emitter path. Conventional current flow is from the collector to the emitter in an npn transistor and vice versa for a pnp transistor. The switching speed of transistors is faster than electromagnetic relays. Consequently, transistor outputs reduce response time (see below).

Transistors are used for switching d.c. loads, whereas solid state devices such as triacs (see Chapter 3) can be used to switch a.c. loads. Figure 2.13 shows ways of connecting d.c. loads to transistor outputs. If the output port uses a light-duty npn transistor, as shown in Fig. 2.13(a), a second external power transistor is required because a light-duty transistor is not capable of switching an output load directly. The switching capability of a light-duty transistor will be rated in milliamperes, 25 mA being typical.

Power transistors can be incorporated internally as shown in Fig. 2.13(b). A typical power transistor is capable of switching several tens of amperes. A heat sink is required when switching large currents, as power transistors get very hot. If a heat sink is not used, the current-carrying capacity is very much reduced (usually to under 1 A). Switching capabilities of transistor outputs are always quoted by the manufacturer in terms of the maximum voltage and current that can be used. For example, 24 V/0.5 A for a typical power transistor.

2.9.4 HIGH-SPEED COUNTER INPUTS AND PULSE CONTROL

The PLC is often required to read high-speed pulses from an input device such as a shaft encoder or produce pulses to drive a stepper motor. PLC ports cannot be used to generate or read high-speed pulses

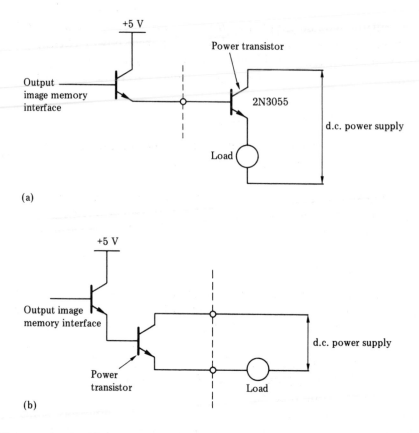

Figure 2.13 Transistor output stage; (a) port requiring external power transistor; (b) port incorporating a power transistor.

as the scan time (which depends on program length, etc.) is a limiting factor. Instead, use is made of interfaces which operate independently of the scan but are able to interrupt it when some action is required.

2.9.5 ANALOGUE PORTS

Many types of transducer produce analogue signals. Variable-speed motor drives are controlled by an analogue speed command signal. Consequently, PLC manufacturers provide ports for handling analogue signals as well as digital. These are based on analogue to digital converters (ADCs) and digital to analogue converters (DACs). Analogue devices are discussed in Chapter 4.

2.9.6 COMMUNICATIONS PORTS

Many PLCs have ports for network communications and for interfacing to a computer. Communications ports are discussed in Chapter 9.

2.10 Response time

The response time of a PLC is the delay between an input being turned on and an output changing state. Delays are due to (a) the mechanical response of an output device such as a relay, (b) the electrical response of an input circuit and (c) the scan update of image memory.

Ladder circuits which feed back the logic states of output relays as inputs can cause a significant response time lag. Consider the ladder shown in Fig. 2.14. Contact $IN1$ energizes output $CR1$ via $CR2$ (i.e. the feedback). Because $CR2$ has to be set by $IN1$ before contact $CR2$ is turned on, it takes a second scan cycle to energize $CR1$. The response lag in this simple example is a single scan cycle.

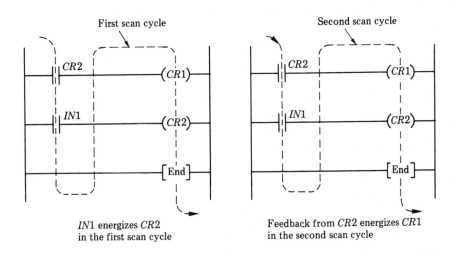

IN1 energizes CR2
in the first scan cycle

Feedback from CR2 energizes CR1
in the second scan cycle

Figure 2.14 Response lag due to the scan.

2.11 Power supplies

The CPU, memory and input/output are electronic components which require power (typically +5 V d.c. and +/− 15 V d.c. at a few milliamperes). A PLC incorporates a power supply for powering internal components and input ports.

Power supplies fall into two categories: linear and switch mode. A linear power supply uses a simple regulator circuit to convert the mains supply to a constant d.c. voltage. A switch-mode power supply uses a high-frequency switching regulator to produce a series of pulses. Averaging the pulses provides a smooth d.c. voltage. The main advantages of a switch-mode power supply are: (a) it is capable of providing a wide range of supply voltages (e.g. +/−24 V d.c., +/−15 V d.c., +/− 5 V d.c., 0 V), (b) switch action makes it highly efficient so that the amount of heat dissipated from the supply is small, and (c) it is compact and lightweight. Because of these advantages the switch-mode power supply is often used in PLCs.

Questions

1. Explain:
   (a) the difference between RAM and ROM;
   (b) the main advantage of EEPROM over EPROM;
   (c) the function of the image memory;
   (d) the function of a memory map.
2. How many bits can a 2K byte RAM hold and how many 8-bit instruction codes does this correspond to?
3. What part of the PLC deciphers the logic of the program?
4. Explain:
   (a) what types of information are stored in data registers;
   (b) the operation of the shift register;
   (c) the term flag;
   (d) the term 'auxiliary relay';
   (e) the term 'phasing error'.
5. Convert the following binary numbers into equivalent octal, hexadecimal and decimal (denary) values:
   (a) 1011
   (b) 11010001
   (c) 1010101010101010
6. What is the BCD code for 321?
7. The power rating of an output relay is 24 watts (watts = volts × amperes). What is the maximum load current that can be applied to the contacts with 24 $V_{r.m.s.}$ and 2.4 $V_{r.m.s.}$?
8. What are the advantages and disadvantages of using:
   (a) relay outputs;
   (b) transistor outputs.

3
Interfacing digital devices

3.1 Introduction

Basically this chapter is a list of control elements used when automating a process or machine. It describes how to interface and connect these elements to digital input/output ports. This will pave the way for the practical applications in Chapters 7 and 8.

As we have seen, digital devices have two distinguishable states, on and off. A simple switch is an obvious example of a digital input. Actuators such as relays, solenoids and motors may be controlled by turning them on and off, and are examples of digital output devices.

Most quantities, for example length, temperature and time, are analogue (i.e. continuous) in nature. One should be clear that the vast majority of so-called digital sensors measure analogue quantities but operate so that they produce a digital output. The encoder is a case in hand. It is a device that measures rotation or linear displacement, which are analogue quantities, but its operation is such that it produces a digital output.

Analogue signals require special attention since they need to be converted into digital quantities so that they may be understood by the CPU. Analogue signals and devices are dealt with in Chapter 4.

3.2 Digital input devices

Provided care is taken over matching voltage and power levels, digital sensors or transducers are connected directly to input ports as either current sourcing or sinking devices. As explained in Chapter 2, a sourcing device is one which provides a current to the input port while a sinking device is one which is able to accept a current from an input port to earth. Some devices such as relays and transistors can be used for sensing and also actuation. Input devices are discussed below.

3.2.1 MECHANICAL SWITCHES

The simplest digital devices are mechanical switches. These are available in various configurations, depending on the number of poles and the number of usable positions, which are referred to as throws.

A simple on/off switch such as a push button has a single-pole single-throw (SPST) configuration. A typical microswitch has a single-pole double-throw (SPDT) configuration. When operated, a microswitch toggles between a normally open and normally closed contact. Figure 3.1 shows typical switch arrangements and ways of connecting them as source inputs.

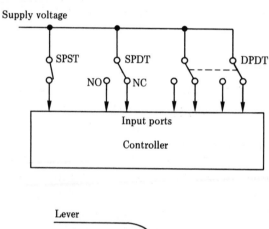

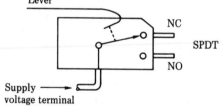

A typical microswitch

Figure 3.1 Switch connections.

3.2.2 TRANSISTOR SWITCHES

Transistors can be used as electrically operated switches. Figure 3.2 shows how to connect pnp and npn transistors to ports that require current sourcing. In both cases, a current applied to the base of the transistor switches on the input port.

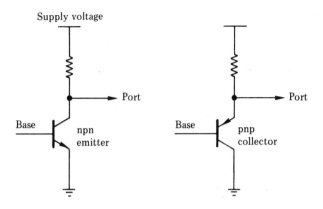

Figure 3.2 Using transistors to switch inputs.

3.2.3 PROXIMITY SWITCHES

Proximity switches, also called proxistors, are non-contact solid-state switches. They switch on and off as an object is moved in and out of a given sensing range. Unlike mechanical switches, they have no moving parts and so there is no mechanical wear. There are three main types of proxistor: inductive, capacitive and optical. Optical proxistors are a form of photoelectric switch and so are explained in Section 3.2.4

Inductive proxistors sense metallic objects. They incorporate an a.c. coil circuit which senses eddy currents inducted by a metal when it is moved close to the coil. The sensing range varies from 0.8 to 15 mm depending on the type of device and whether the metal to be detected is ferrous or non-ferrous.

Capacitive proxistors switch when they sense a change in dielectric constant (permittivity) brought about when a material is moved close to the device. As all materials have a dielectric constant they are not restricted to sensing only metals, as is the case for inductive proxistors. Capacitive proxistors have a similar sensing range to inductive devices.

The output switching stages of proxistors are generally pnp and npn transistors as illustrated in Fig. 3.3. Npn and pnp proxistors are used as source and sink devices respectively. Proxistor sensing rates vary between 10 and 30 Hz. A proxistor with a sensing rate of 10 Hz is able to sense ten objects per second. Many PLC manufacturers provide high-speed counter inputs for proxistors so that phasing errors do not occur in high-speed sensing applications.

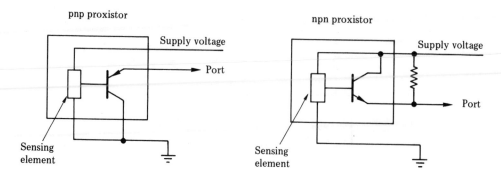

Figure 3.3   Proxistors.

3.2.4 PHOTOELECTRIC SWITCHES

Photoelectric switches are opto-electronic devices consisting of a light source (transmitter) and light receiver. They are used to detect an interruption or reflection of a light beam. There are three types of photoelectric switch, namely, the through beam, retro-reflective and light scanner devices (see Fig. 3.4):

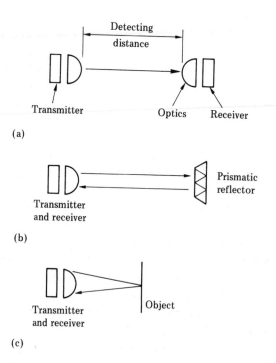

Figure 3.4 Types of photoelectric switch: (a) through beam; (b) retro-reflective; (c) light scanner.

1. *Through-beam* This type of photoelectric switch consists of two separate units; the transmitter and receiver. A light beam generated from the transmitter is directed at the receiver. An object which breaks the light beam triggers a switch in the receiver.

   Through-beam devices can be used over long distances (up to 100 m). However, poor alignment of the transmitter and receiver can lead to problems. Clearly such devices cannot be used to detect transparent objects.

2. *Retro-reflective* In this type of photoelectric switch the transmitter and receiver are contained in the same body. A light beam generated from the transmitter is directed at a prismatic reflector which returns the light beam to the receiver. Breaking the light beam triggers a switch in the receiver.

   It is important to use a prismatic reflector with this type of device and not an ordinary mirror. A prismatic reflector reflects a light beam parallel to the one it receives – a process which is called retro-reflection (hence the name retro-reflective photoelectric switches). Retro-reflective photoelectric switches have a range up to 5 m, and polarizing filters may be used to prevent stray reflections causing false triggering.

3. *Light scanners* This type of photoelectric switch is used to detect the surface reflection from an object. Like the retro-reflective type of device the transmitter and receiver are contained within the same body. However, with light scanners the object surface is used as the reflector. In effect this type of device is an optical proxistor.

   Generally the detecting distance of light-scanning photoelectric switches is short (up to 2 m). The distance depends on the object's surface colour, texture and reflectivity. It also depends on background lighting.

3.2.5 REED SWITCHES/RELAYS

Reed switches are magnetically operated switches. They consist of two overlapping, but not touching, nickel–iron contacts sealed in an inert atmosphere by a glass tube. A magnetic field causes the contacts to close. The magnetic field can be produced by a permanent magnet or coil. If a coil is used the device is called a Reed relay. Reed devices have high operating speed and low contact resistance. The symbol for a Reed switch is shown in Fig. 3.5.

Reed switches are often mounted on the outside of pneumatic cylinders to detect the presence of a magnetic plunger. Figure 3.5 illustrates how this is done. When the magnetic plunger of the piston is moved to either end of the cylinder one of the Reed switches turns on.

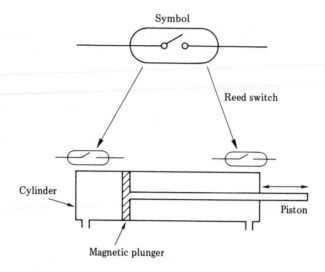

Figure 3.5 Reed switch.

3.2.6 ENCODERS

Encoders are devices which convert rotation and linear displacement into digital signals. Linear encoders measure linear displacement. Shaft encoders measure rotation. Encoders are classified as to whether they are incremental or absolute. Incremental encoders measure position relative to some reference point. Absolute encoders produce a coded number which corresponds to a position.

An example of an incremental shaft encoder is shown in Fig. 3.6. It consists of a disc with evenly spaced clear and dark markings sandwiched between a light source and a photocell. When the shaft is turned, the disc rotates and chops the light beam so that the photocell produces a series of pulses. Position is measured relative to a reference point by counting pulses generated by the photocell as the disc rotates from the reference point.

Many incremental shaft encoders use two photocells positioned so that they produce two pulses that are 90° out of phase with each other. The logic state of the second pulse with respect to the first can be used to determine direction of rotation. If the shaft is driven by a motor the pulse rate is proportional to the motor's speed.

An example of an absolute shaft encoder is shown in Fig. 3.7. The encoder consists of a disc sandwiched between a light source and three photocells. The disc is shaded using clear and dark markings such that a unique code is produced by the photocells for each shaft position. Discs are normally shaded with a code, called Gray code, which changes

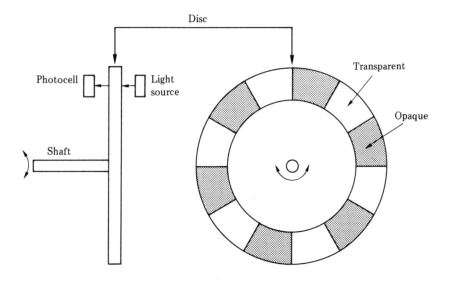

Figure 3.6 Incremental shaft encoder

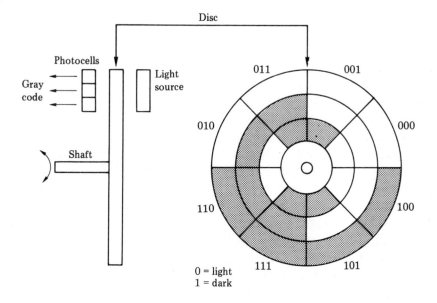

Figure 3.7 Absolute 3-bit Gray-coded shaft encoder.

by a single bit between successive positions. Table 3.1 compares Gray code with binary code.

Linear encoders use a slide shaded with clear and dark markings rather than a disc, otherwise their operation is similar.

Table 3.1 Binary and Gray codes

Binary	Gray
0000	0000
0001	0001
0010	0011
0011	0010
0100	0110
0101	0111
0110	0101
0111	0100

3.2.7 TEMPERATURE SWITCH

Temperature switches are semiconductor devices which exhibit a rapid change in resistance around a specific transition temperature. A typical device maintains a high resistance of, say, 100 kohm, until heated to the transition temperature of 75°C, whereupon the resistance drops to 100 ohm. When the device exhibits a high resistance it is effectively 'off' as very little current is drawn. The device turns on when the resistance between its terminals is low. Temperature switches are used to protect motors, transformers and power semiconductors from overheating. They can also be used as a temperature alarm input.

Analogue temperature sensors such as thermocouples and semiconductor diode sensors in conjunction with a comparator amplifier can be used as 'digital' temperature switches. Figure 3.8 illustrates how it is possible to use a temperature diode as a thermal switch. Temperature diodes pass current in proportion to temperature. A typical device has a temperature coefficient of 1 μA/K. The circuit in Fig. 3.8 converts the sensor current into voltage by using resistor $R1$. This voltage is amplified by $A1$ and fed into the comparator amplifier $C1$. The comparator produces a 'logic 1' output when the temperature voltage input (Vin) is equal to or greater than the set point or reference input (Vref) and a 'logic 0' otherwise. The comparator output switches transistor $T1$ on and off, and this in turn switches the input port on and off. The switching temperature is adjusted via resistor R_V.

3.2.8 PRESSURE SWITCHES

Pressure switches are devices which turn on and off at a certain transition pressure. Operation is such that a diaphragm moves in proportion to fluid pressure and operates a microswitch. Consequently, pressure switches are connected to input ports as standard switch inputs.

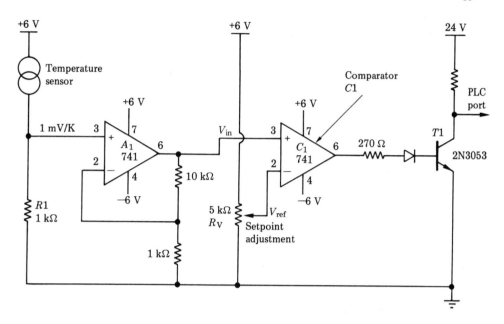

Figure 3.8   Temperature switch circuit.

3.3   Digital output devices

Digital output ports control actuators which in turn control process variables or machine actions. Typical actuators are solenoids, contactors, solid state relays and motors.

3.3.1   SOLENOIDS

A solenoid is an electromagnetic device which converts an electrical signal into mechanical movement. Essentially it consists of a coil and moving armature. Energising the coil causes the armature to move. The distance moved is referred to as the stroke. Typical stroke lengths range from 4 to 20 mm depending on the device and applied force. Typical operating voltages are 12 V d.c., 24 V d.c./a.c. and 240 V a.c. with coil current being rated at a few amperes (d.c.). Fig. 3.9 shows how to connect a 24 V a.c. solenoid to a relay-driven output port. A snubber (*RC* network) is connected across the relay contacts to suppress noise (see Section 3.4) which is caused when switching a reactive load. It also improves the contact life of the relay.

3.3.2   CONTACTOR

A contactor is a heavy-current switching device which is actuated by either mechanical or electrical means. An electrically operated contactor is illustrated in Fig. 3.10. It consists of a solenoid, magnetic core

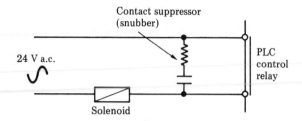

Figure 3.9 Connecting a solenoid to a control relay.

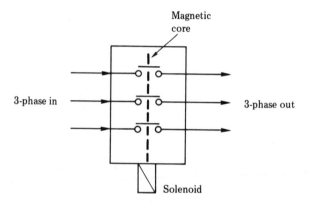

Figure 3.10 Three-phase contactor.

and contacts. When the solenoid is energized the magnetic core is attracted to it, so closing heavy duty contracts. The solenoid side of the contactor can be controlled from an output port. Contactors are used to switch single and three-phase supplies and are widely used in industrial control.

3.3.3 SOLID-STATE RELAY

A solid-state relay performs the same function as a traditional relay but has no moving parts. It is basically an optically isolated triac. As shown in Fig. 3.11 a triac is two back-to-back silicon diodes which are switched into conduction by a third electrode called the gate. As a solid-state relay is optically isolated the gate may be thought of as being triggered by an LED.

Solid-state relays are used for switching mains and often incorporate a zero crossing circuit (ZCC). This monitors the mains cycle so that the can be turned on at a zero crossing point. Switching at a zero crossing point prevents high-frequency noise being generated. However, on and off delays do occur as a result of waiting for the next zero crossing point (see Fig. 3.11).

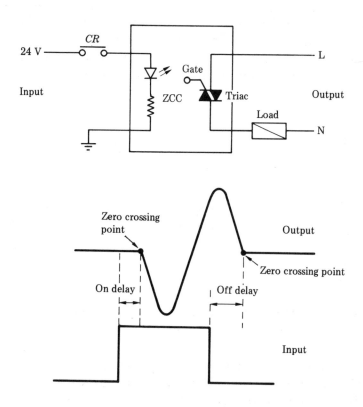

Figure 3.11 Solid-state relay which switches at a zero crossing (ZC) point.

3.3.4 D.C. MOTORS

All non-linear motors are constructed so that they have a central rotor which is free to rotate within a stator. A d.c. motor consists of an armature winding mounted on the rotor, a field winding mounted on the stator and a commutator/brush arrangement for changing the polarity of the d.c. current to the armature as it rotates. The stator may be a permanent magnet instead of a field winding. A torque T is produced from the interaction of the armature current I_a and magnetic field flux Φ of the stator, i.e.

$$T = K_t \, \Phi \, I_a,$$

where K_t is the torque constant given in manufacturers data.

D.C. motors are described by the way in which the d.c. supply is connected to the armature and field windings (see Fig. 3.12). The series d.c. motor has the armature and field windings connected in series. The shunt d.c. motor has both the armature and field windings connected in parallel. The compound d.c. motor has both shunt and series connections. Many d.c. motors have four terminals – two for the armature windings and two for the field windings – and so can be configured as either series or shunt. A d.c. motor with only two terminals has fixed configuration as the windings are connected internally.

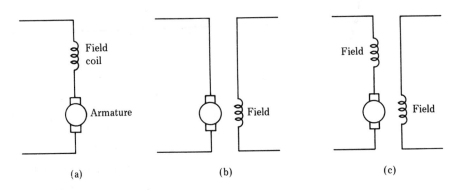

(a)   (b)   (c)

Figure 3.12   Direct current motor configurations; (a) series; (b) shunt; (c) compound.

A d.c. motor can be turned on and off by connecting an output port in series with the armature (see Fig. 3.13). The purpose of the diode in the circuit is to dissipate induced current due to back e.m.f. A reverse-connected diode is always used when switching an inductive load such as a motor winding or relay coil.

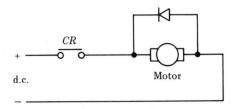

Figure 3.13   On/off control of a d.c. motor.

Many industrial applications require a PLC to switch a d.c. motor on/off and also to be able to change the direction of the motor. The simplest way of doing this is to use relays as shown in Fig. 3.14. Relays *CR*1 and *CR*2 in the circuit change over the positive and negative sides of the power supply and so reverse the direction of the motor. Relay

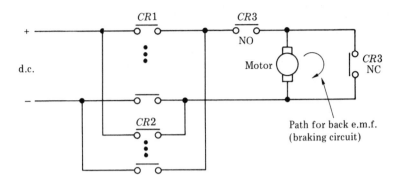

Figure 3.14 On/off and direction control of a d.c. motor.

*CR*3 stops and starts the motor and provides a path for the current induced in the motor windings due to back e.m.f. Without this path, the back e.m.f. current would cause the motor to overrun when stopped. This path acts as a braking circuit. Note that if the power to *CR*3 fails the motor is automatically stopped.

A d.c. motor draws peak current when power is first applied (this is called the start-up current) and when the mechanical load becomes so great the motor stalls (this is called the stall current). Start-up and stall current are roughly equal and should never be allowed to exceed the maximum current rating of switching relays or transistors. Stall current is calculated as V_s/R where V_s is the motor supply voltage and R the winding resistance presented to the motor supply. The winding resistance is easily measured with an ohm-meter.

3.3.5 A.C. MOTORS

There are three types of a.c. motors: (a) the universal motor, (b) the single-phase induction motor, and (c) the three-phase induction motor. The universal motor is in fact a d.c. series motor. It is termed universal because it can be run from either a d.c. or a single-phase a.c. supply. The single-phase a.c. motor is not self-starting but otherwise operates like a three-phase motor.

The three-phase induction motor consists of stator windings and a cage-like rotor which resembles a squirrel cage (hence the name squirrel-cage induction motor). They work on the principle of electro-magnetic induction according to Faraday's law. Three-phase currents are applied to stator windings and cause a rotating magnetic flux which induces currents in the cage rotor. The induced rotor currents and stator flux interact to produce torque. The rotor attempts to follow the rotating stator flux. The speed difference between the rotating flux and rotor is called the slip.

Three-phase induction motors can be switched on and off using a contactor controlled from an output port. Alternatively, three solid-state relays can be used to switch each of the three phases. A fourth solid-state relay can be used to reverse two of the phases and so reverse motor direction. Figure 3.15 shows ways of connecting a three-phase motor to relay-based output ports.

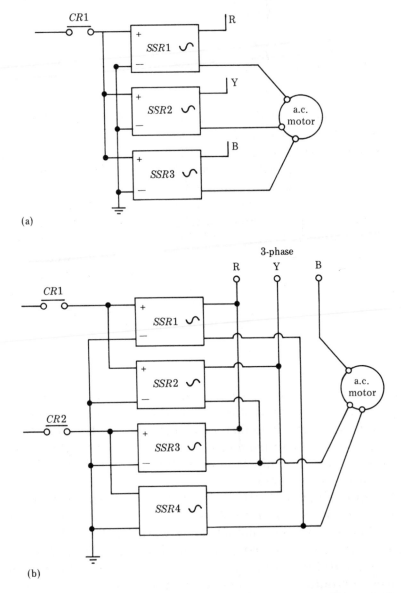

(a)

(b)

Figure 3.15   (a) On/off control of a three-phase motor; (b) on/off and directional control of a three-phase motor.

3.3.6 STEPPER MOTORS

Stepper motors are driven by digital pulses. Each pulse causes the shaft of the motor to rotate through a discrete step angle. The step angle of a particular motor is determined by its construction but 1.8° is typical. Stepper motors are ideally suited to positioning applications because step-angle errors are small and do not accumulate as a number of steps are made. This means a position feedback loop is not required.

Three types of stepper motor are available: (a) variable reluctance, (b) permanent magnet, and (c) hybrid. A variable-reluctance stepper motor is capable of stepping at high speeds but has no detent torque (i.e. no resistance to movement when the windings are not energized). A permanent magnet stepper can be used in applications which require holding torque when the motor is unpowered. Hybrid motors are capable of operating at high speed and maintain a holding torque when their power supply is removed.

Stepper motors are controlled by drives which require clock and direction inputs. The motor steps each time the clock input is pulsed. Motor speed is controlled by varying the clock pulse rate. The step rate is severely limited by scan time if a standard PLC output port is used to provide the clock signals. To get round this problem, many PLCs have special NC (numerical control) ports which allow the user to vary pulse rate through program functions. These incorporate an electronic clock circuit for producing the step pulses.

3.4 Noise

Noise is an unwanted component in a signal. PLCs are designed to withstand noisy power supply and input signals. Even so, high levels of noise can affect the working of a PLC causing problems such as false input triggering and power failure.

Examples of noise are: voltage spikes, switching transients, radio-frequency interference, mains-frequency hum, d.c. drift and thermal and semiconductor noise (i.e. white and flicker $1/f$ noise). Some examples of noisy signals are illustrated in Fig. 3.16. Voltage spikes and transients are normally caused by high-voltage switching circuits. Radio-frequency interference is caused by d.c. motors, arc welders and triac switching circuits. Mains hum is due to mains interference. Drop outs (i.e. sudden reduction in supply voltage) can occur when heavy current motors are started. Thermal and semiconductor noise is generated by resistors and amplifiers and can cause problems in applications involving small-signal recovery. D.C. drift can also be problematic in badly designed amplifiers.

When switching a.c. devices snubbers (contact suppressors) should be used as these reduce switching transients. A snubber is a resistor and

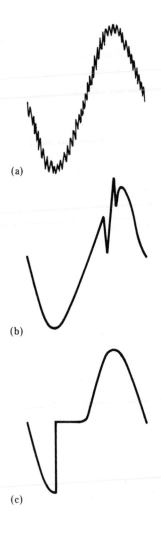

(a)

(b)

(c)

Figure 3.16 Examples of noise; (a) radio-frequency interference; (b) spikes, (c) drop out.

capacitor connected in series (100 ohm and 0.1µF being typical values). The electrical snubber is analogous to the mechanical snubber, which is a miniature shock absorber used to damp out mechanical oscillations.

Metal oxide varistors (MOVs) are voltage-dependent resistors. They are used to absorb voltage spikes and are connected across input devices. Isolating transformers can be used to separate a noise-sensitive device from a noise source. Mains filters reject 50 Hz interference. Grounding all equipment at a single point prevents mains interference due to ground loops. A noisy mains earth line should never be connected to a PLC. All cables should be shielded to prevent radio frequency and mains pick-up.

Questions

1. Show how an SPDT microswitch, an npn proxistor and a pnp proxistor are connected as source inputs to ports which use 24 V/0 V logic levels.
2. Explain the operation of an incremental shaft encoder.
3. What types of heavy-duty devices can be used for secondary switching?
4. Show how secondary relays are used to turn a d.c. motor on and off and also change its direction.
5. What is a snubber?

4

Interfacing analogue devices

4.1 Introduction

The devices we have looked at so far have all been digital, having two discrete levels, on and off. Analogue signals are quite different. These are continuous signals which vary in level.

PLCs are digital devices. To handle analogue signals special interfaces based on analogue to digital converters (ADCs), digital to analogue converters (DACs), multiplexers and de-multiplexers are required. These are discussed below.

4.2 Digital to analogue converter

A digital to analogue converter (DAC) produces an analogue output from a digital input (see Fig. 4.1). In all types of DAC the analogue voltage is produced from a reference voltage (V_{ref}). Binary code is input into the DAC and determines what fraction of V_{ref} is presented at the output. The output from a DAC is not truly continuous but rather a series of discrete voltage levels.

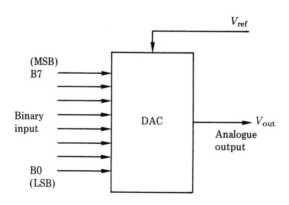

Figure 4.1   Eight-bit DAC.

For example, the 8-bit DAC shown in Fig. 4.1 has an output given as

$$V_{out} = V_{ref} \left(\frac{B7}{2} + \frac{B6}{4} + \frac{B5}{8} + \frac{B4}{16} + \frac{B3}{32} + \frac{B2}{64} + \frac{B1}{128} + \frac{B0}{256} \right) \qquad (4.1)$$

where the bits B7 to B0 can take values 0 or 1 and are the binary inputs. B7 is the most significant bit (MSB) and B0 the least significant bit (LSB).

Consider an 8-bit DAC with a reference voltage V_{ref} of 10 V. The binary input of 00000001 generates the smallest discrete output, i.e. 10/256 volts. The next discrete output is 10/128 volts, generated from the binary code 00000010. Clearly 256 discrete analogue levels (referred to as quantization levels) can be produced from the binary input. The voltage resolution of an N-bit DAC is calculated by dividing the maximum operating voltage by $2^N - 1$. The factor $2^N - 1$ represents the number of steps between quantization levels. An 8-bit DAC with a reference voltage of 10 V has a resolution of 10/255.

The speed of a DAC is determined by how long it takes to settle to a stable value after a change in input. This is specified as the settling time. The other main parameters of a DAC are linearity and accuracy. Linearity is a measure of the deviation from a straight line of output voltage plotted against binary input. Accuracy is the variation between the DAC's actual output and the intended one.

The operating principle of a DAC is illustrated in Fig. 4.2. DACs use a set of binary weighted resistors switched by the binary input to generate an analogue output from V_{ref}. The highest weighted resistor generates the smallest discrete value at the analogue output.

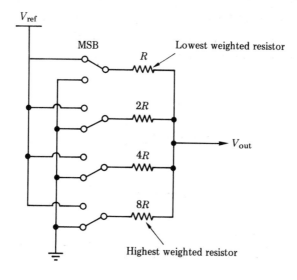

Figure 4.2   Operating principle of a DAC.

4.3 Analogue to digital converter

The analogue to digital converter (ADC) produces a digital output from an analogue input (see Fig. 4.3). ADCs incorporate start convert (SC) and end of convert (EOC) connections. When the start convert signal is pulsed the ADC converts the analogue input at that time into an equivalent digital value. The ADC then produces an end of convert signal to indicate that the conversion has finished.

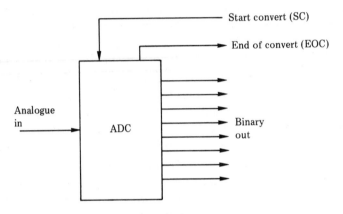

Figure 4.3 Eight-bit ADC.

The simplest type of ADC makes use of a DAC and a comparator as shown in Fig. 4.4. Digital data from a counter is fed into the DAC and generates an analogue voltage which is compared with the incoming analogue voltage which is to be converted. When both signals match, the comparator amplifier generates a logic 1 to indicate that conversion has finished (i.e. the end of convert signal). The digital value input to the DAC at that time represents the analogue input.

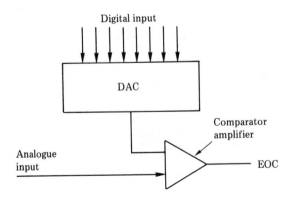

Figure 4.4 Operating principle of a comparator type ADC.

The main parameters of ADCs are again resolution, accuracy, linearity and speed. The comments already made concerning resolution, accuracy and linearity of DACs apply to ADCs. Note that for an 8-bit ADC equation (4.1) works in reverse. Concerning operating speed, ADCs are generally slower than DACs because the process involves comparing one signal with another. Successive approximation of the input value rather than ramping the DAC from a counter speeds up the conversion process. For high speed, so called 'flash' converters are used.

4.4 Multiplexer

A multiplexer allows several signal carrying channels to share a single line. A block diagram of a multiplexer is illustrated in Fig. 4.5. This shows that each input channel may be connected to the output line when one of a bank of switches inside the multiplexer is turned on. In practice the bank of switches shown in Fig. 4.5 is a bank of transistor

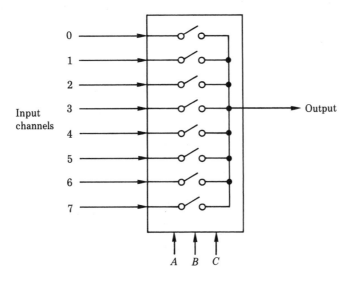

C	B	A	Channel selected
0	0	0	0
0	0	1	1
0	1	0	2
0	1	1	3
1	0	0	4
1	0	1	5
1	1	0	6
1	1	1	7

Figure 4.5 Multiplexer.

switches controlled by the lines A, B and C. A binary code placed on the lines labelled A, B and C determines which of the channels is switched through to the output. De-multiplexers are multiplexers that work in reverse.

4.5   Interfacing

The general rule when interfacing analogue signals is to match voltage levels and ensure that the impedance of the sourcing circuit is less than or equal to that of its load circuit. Impedance matching is essential for the optimum power transfer to the load circuit. To match voltage levels you may have to reduce or amplify a voltage level or perhaps convert a bipolar voltage into a unipolar voltage. To meet the impedance criteria you may have to use a circuit for changing impedance. Some simple interfacing circuits are discussed below.

A potential divider is a series arrangement of resistors (see Fig. 4.6) which is used to reduce the voltage level of an input signal. The output voltage V_o is always less than the input voltage V_i by an amount given by the formula

$$V_o = V_i R_2/(R_1 + R_2).$$

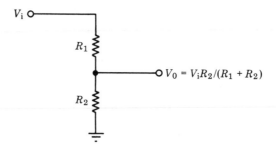

Figure 4.6   Potential divider.

Inverting and non-inverting amplifiers may be used to increase a voltage level. Figure 4.7 shows circuits for inverting and non-inverting amplifiers which make use of a 741 operational amplifier (abbreviated to op-amp). The ratio of the output voltage V_o to the input voltage V_i is called the gain of the amplifier. The inverting amplifier produces a negative output because it has a gain of $-R_2/R_1$. The non-inverting amplifier has a gain of $(R_2+R_1)/R_1$. The resistor R_i in these circuits ensures that both inputs to the operational amplifier see the same resistance to earth and is calculated from the formula

$$R_i = R_2 R_1/(R_1 + R_2).$$

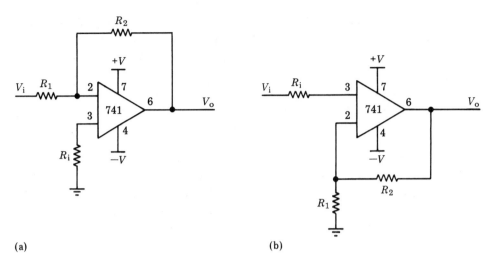

Figure 4.7   (a) Inverting and (b) non-inverting amplifiers.

A summing amplifier may be used to convert a bipolar voltage into a unipolar voltage. Figure 4.8 shows a circuit for a summing amplifier which makes use of a 741 operational amplifier. The output voltage V_o is given by

$$V_o = -(V_1 + V_2).$$

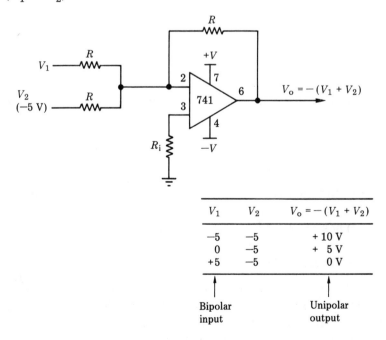

V_1	V_2	$V_o = -(V_1 + V_2)$
−5	−5	+ 10 V
0	−5	+ 5 V
+5	−5	0 V

Bipolar input   Unipolar output

Figure 4.8   Summing amplifier.

If the amplifier is to convert a bipolar voltage into a unipolar voltage, one of the inputs must be held at an appropriate negative voltage. For example, if V_2 is held at -5 V, then summing action ensures that a bipolar voltage of -5 V through to $+5$ V on V_1 is converted into a unipolar voltage ranging from 10 V to 0 V (see the table of Fig. 4.8).

The emitter follower and voltage follower (a unity gain non-inverting amplifier) are circuits which provide an impedance match between a source and load. The source might be an amplifier or DAC output and the load the analogue input to a servo drive (servo drives are described below). Figure 4.9 shows circuits for the emitter follower and voltage follower. In these circuits, the output voltage follows the input voltage but the input impedance is much larger than the output impedance. Thus, they lower the impedance of the source.

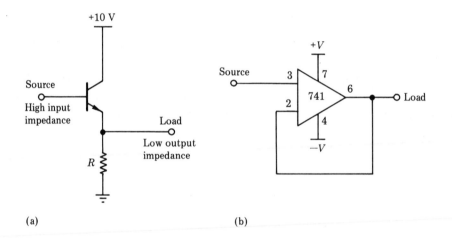

Figure 4.9   Impedance changing circuits: (a) emitter follower; (b) voltage follower.

4.6   Analogue transducers

Physical quantities such as displacement, temperature and strain are converted into analogue voltage or current by transducers. Some common types of transducer are described below.

4.6.1   POTENTIOMETERS

The simplest way of producing an analogue input to an ADC is to use a potentiometer circuit such as that shown in Fig. 4.10. The position of the wiper terminal of the potentiometer is converted into a voltage signal. Linear and rotary potentiometers may be used as low-cost position transducers.

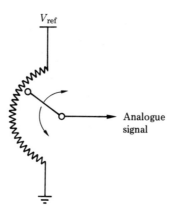

Figure 4.10   Rotary position potentiometer.

4.6.2   LVDT

The linear variable differential transformer or LVDT is a displacement transducer. It consists of a nickel-iron rod which is free to move through primary and secondary coils. The secondary winding is connected to a phase-sensitive rectifier (PSR). The basic arrangement   is illustrated in Fig. 4.11.

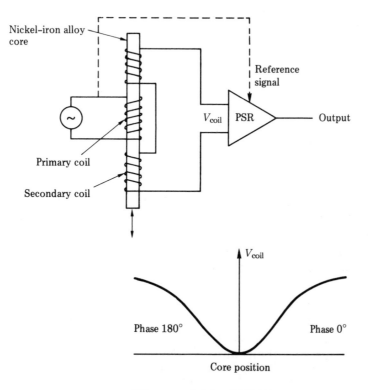

Figure 4.11   LVDT.

The primary coil is fed alternating current so that voltages are induced in the two halves of the secondary coil. Moving the rod up and down changes the phase and voltage in the secondary windings. The PSR amplifier is used to translate the phase and voltage changes in the secondary coil into a unipolar or bipolar output voltage. The output of a typical unipolar LVDT may be fed directly into an ADC.

4.6.3   THERMOCOUPLE

A thermocouple consists of two dissimilar wires which are arranged as shown in Fig. 4.12. Voltage is produced by thermoelectric effects as the hot junction is heated. Thermocouple types which conform to British standards are designated letters. These letters determine the metals used in the thermocouple junction (see Fig. 4.12).

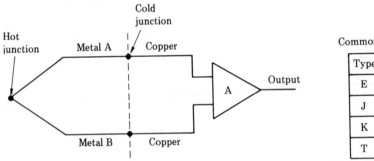

Common types of thermocouple

Type	Metal A/metal B
E	Chromel/constantan
J	Iron/constantan
K	Chromel/alumel
T	Copper/constantan

Figure 4.12   Thermocouple.

Thermocouples are non-linear devices which means that their output voltage is not proportional to temperature. Indeed, a thermocouple will be supplied with a calibration table which must always be referred to when converting an output voltage into temperature.

The voltage produced by a thermocouple needs to be amplified before it can be fed into an ADC unit. A simple thermocouple amplifier circuit, based on the 741 operational amplifier, is illustrated in Fig. 4.13.

Semiconductor temperature diodes are available which pass current when heated. They operate over a smaller temperature range compared to thermocouples but their output may often be treated as linear. Temperature coefficients vary but 1 μA/K is typical.

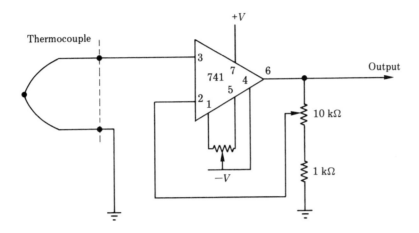

Figure 4.13   Thermocouple amplifier.

4.6.4 STRAIN GAUGE

A strain gauge is a device which changes resistance when stretched or compressed. The relationship between the change in resistance ($\Delta R/R$) and corresponding change in strain (i.e. length change $\Delta L/L$) is given as

$$G = \frac{\Delta R/R}{\Delta L/L}$$

(4.2)

where G is called the gauge factor.

The gauge factor G is about 2 for metal alloy strain gauges and about 100 for semiconductor strain gauges. Although it is possible to measure the strain directly using equation (4.2), it is normal practice to use a balancing bridge circuit of the type shown in Fig. 4.14. The analogue output of the bridge is nulled using the variable resistance R_V when no strain is applied to the gauge. When the gauge is strained a voltage in the bridge circuit is observed because the bridge is no longer balanced. This voltage is usually amplified and fed back into an ADC so that it may be compared with a calibrated value. Because of their large gauge factor, semiconductor strain gauges produce much larger signals compared with metal types. However, they are more sensitive to temperature variation.

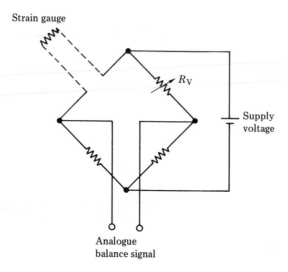

Figure 4.14   Strain gauge bridge.

4.7   Servo drives

Servo drives are generally used to control the speed or shaft position of a motor. The output of a drive is usually controlled by an analogue input voltage which may be supplied by the PLC's DAC. Servo drives are either open-loop or closed-loop systems.

4.7.1   OPEN-LOOP AND CLOSED-LOOP SERVO DRIVES

A comparison of open-loop and closed-loop servo drive systems is shown in Fig. 4.15. In both cases, θ_o the motor output (i.e. either speed or position), is controlled by varying the electrical input signal θ_i. Open-loop servo drives do not check to see if the actual output is at the desired value set by θ_i. Closed-loop servo drives, by using feedback transducers, compare the actual output of the motor with the desired value. A comparator is used to obtain an error signal E by subtracting the input signal from the output signal ($E = \theta_i - \theta_o$). The error drives the amplifier and provides corrective action to ensure that the output is maintained if disturbed.

   Figure 4.16 illustrates closed-loop d.c. servo drives used for position and speed control. Speed control requires velocity feedback only. Position control requires both position and velocity feedback (velocity feedback provides damping). A potentiometer with its wiper driven by the motor's shaft can be used as a position feedback transducer. A tachogenerator can be used as a speed feedback transducer. A tacho-generator is essentially a motor used as a generator. The generated

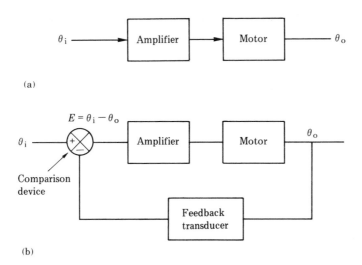

(a)

$$E = \theta_i - \theta_o$$

(b)

Figure 4.15 A comparison of (a) open-loop and (b) closed-loop servo drives.

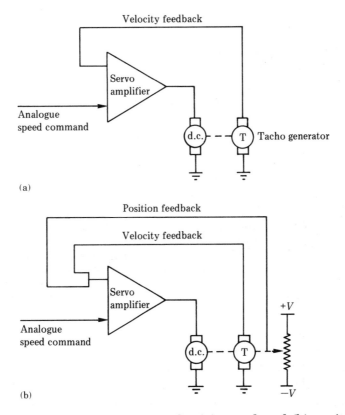

Figure 4.16 Basic arrangements for (a) speed and (b) position d.c. servos.

voltage is proportional to speed. Shaft encoders are also used as position and velocity feedback transducers.

Closed-loop servo drives, although more accurate than open-loop systems, may become unstable if, for example, the amplifier gain is set too high. When unstable the motor oscillates continuously.

4.7.2 CONVERTER DRIVES

A d.c. converter drive converts a single or three-phase supply into a controllable d.c. supply. They are used to control the speed of a d.c. motor by varying the armature current.

4.7.3 INVERTER DRIVES

An inverter drive electronically generates a variable-frequency a.c. supply from a d.c. input. They are used to control the speed of an a.c. induction motor. The speed N of an a.c. induction motor is determined from

$$N = \frac{60F}{P}$$

where F is the supply frequency and P is the number of stator pole pairs. Since the number of pole pairs of a motor is fixed and the supply voltage remains constant, speed can be controlled by varying the supply frequency.

Questions

1. Explain the operation of a DAC.
2. Explain the operation of an ADC.
3. What is the analogue output of an 8-bit DAC having a reference voltage of 10 V if the binary number 10000000 is input?
4. Draw in block-diagram form an arrangement which allows eight analogue inputs to be connected to a PLC having a single ADC input.
5. Show how a 0–10 V signal produced from an analogue transducer may be interfaced to an ADC which accepts an input in the range 0–2.55 V.
6. Show how a PLC may be interfaced to a variable-speed motor drive.

5
Ladder programming

5.1 Introduction

This chapter is basically a programmer's guide to the ladder-diagram programming method. It builds on the work covered in Chapter 1.

5.2 Ladder structure

In Chapter 1 we described a ladder diagram as a circuit consisting of contacts and coils. The PLC scanning process is such that conduction through contacts is from left to right, i.e. from the 24 V bus line. Contacts are always horizontal branches as the scanning process does not allow for vertical branches.

Ladder circuits comprising only contacts and coils are limited to switching operations. To allow PLCs to handle more advanced control tasks manufacturers incorporate a set of special functions. Function sets vary from PLC to PLC but invariably include counters, timers and data-handling routines. Functions are drawn either as special coils or as block elements (see Fig. 5.1). Thus, ladder diagrams comprise contacts, coils and special functional coils or blocks. The following notation is used for the ladder diagrams developed in this chapter:

IN = input contact
CR = control relay
AR = auxiliary relay
T = timer
C = counter

5.3 Multiple outputs

In ladder diagramming, it is possible to connect more than one coil to a contact. This allows a sequencing ladder of the type shown in Fig. 5.2 to be arranged. In this example, $CR1$, $CR2$ and $CR3$ are energized one

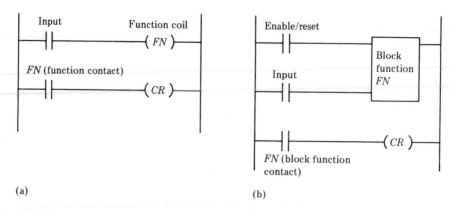

(a)   (b)

Figure 5.1   Ladder functions; (a) function coil; (b) block form.

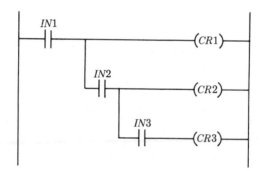

Figure 5.2   Multiple outputs.

after the other provided that the input contacts are turned on in the sequence $IN1$, $IN2$, $IN3$.

5.4 Latching

A latch circuit is used to hold a coil energized even if the input contact which energized it subsequently turns off. An example of a latch circuit is shown in Fig. 5.3. When contact $IN1$ is momentarily turned on, control relay $CR1$ is energized. The logic state of $CR1$ is fed back as an input to keep itself energized when $IN1$ turns off. The latch circuit is broken (i.e. reset) by momentarily turning on $IN2$ so that it opens. The ability to latch on to an input is possible because ladder programming allows an output (i.e. a coil) to be fed back as an input (i.e. a contact).

It is often necessary to retain an output latch during power failure. Battery-backed auxiliary relays can be used for this purpose. An example of a retained latch is shown in Fig. 5.4. The coils $CR1$ and $AR1$

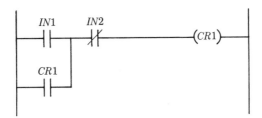

Figure 5.3 Latch.

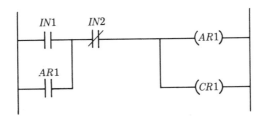

Figure 5.4 Retained latch using a battery-backed auxiliary relay.

(auxiliary relay one) are energized by contact $IN1$. $AR1$ is fed back to latch itself and $CR1$. If $AR1$ is battery backed it remains closed during a power failure. Thus, when power is restored, the coils still remain latched.

Some PLCs use functions for latching and unlatching coils so that the programmer does not have to draw a latching circuit of the types shown in Figs. 5.3 and 5.4.

5.5 Timers

Timers count seconds or fractions of seconds by using the internal CPU clock. They permit delay of the time of certain control operations. The preset value of a timer is the delay period required and is typically set in the range 0.1–999 s in steps of 0.1 s.

Various types of timer are used. Examples include the following.

1. *Pulse timers*   When activated this type of timer generates a single pulse of fixed pulse width. There are two types of pulse that can be generated, positive-going and negative-going, as shown in Fig. 5.5. This figure defines the terms pulse width, leading edge and trailing edge.
2. *Delay-on timers*   When started, this type of timer waits for a fixed delay period before turning on. The timing diagram for a delay-on timer is illustrated in Fig. 5.6. All PLCs use delay-on timers.

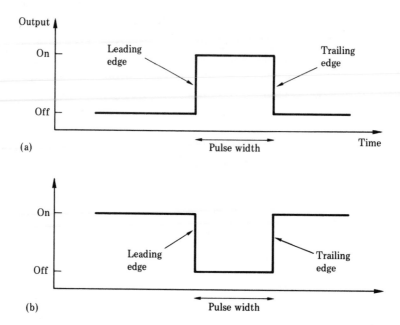

Figure 5.5   Two types of pulse; (a) positive-going; (b) negative-going.

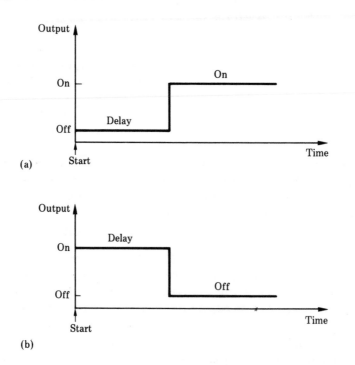

Figure 5.6   Timing diagrams for delay-on and delay-off timers.

3. *Delay-off timers* When started, this type of timer turns on for a fixed delay period before turning off. The timing diagram for a delay-off timer is illustrated in Fig. 5.6.

4. *Repeating timers* Repeating timers allow coils to be repeatedly turned on and off at regular intervals. They are also known as cyclic timers.

Figure 5.7 shows an example of a ladder circuit which uses a delay-on timer labelled *T*1. The timer is represented as a coil. The associated constant represents the delay period of the timer in seconds. In this example, the delay period is preset to 5 s. The timer's normally open

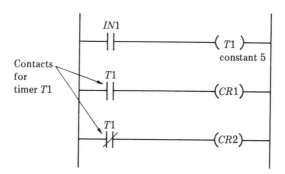

Figure 5.7 Timer circuit.

(NO) and normally closed (NC) contacts are connected to *CR*1 and *CR*2 respectively. Initially, *CR*1 is de-energized and *CR*2 is energized. When the contact labelled *IN*1 is turned on the timer starts measuring time. When 5 s have elapsed the timer turns on and its contacts energize *CR*1 and de-energize *CR*2. The timer is reset to the initial state by turning off *IN*1. A timing diagram for a delay-on timer is shown in Fig. 5.8.

Timers may be cascaded (i.e. linked together) to give large delay times. Fig. 5.9 shows an example of cascaded timers. Timer *T*1 feeds timer *T*2. The total time delay is the sum of the two preset values – in this example, 120 s.

5.6 Counters

Counters are used for counting a specified number of contact operations. The ladder diagram representation of a counter involves two coils, one to count input pulses and one to reset the counter. All counters have an associated constant which represents the count required. This constant is called the preset value.

There are two types of counter: the up-counter and the down-counter. Up-counters (UCs) count up to the preset value. When the preset value

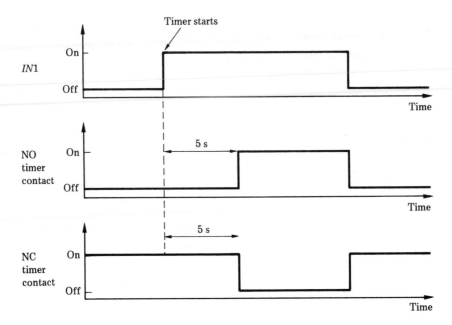

Figure 5.8   Timing diagram for the timer circuit shown in Fig. 5.7.

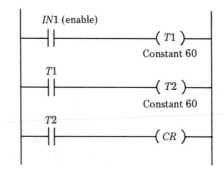

Figure 5.9   Cascading timers.

is reached the counter's contacts change state. Down-counters (DCs) count down from the preset value to zero. When zero is reached the counter's contacts change state.

Fig. 5.10 shows an example of a ladder circuit which uses a counter. The counter is represented by two coils labelled $C1$. One coil is used to count the switch operations of contact $IN1$ and the other is used to reset the counter. The preset count value is equal to three. The counter's normally open (NO) and normally closed (NC) contacts are connected to $CR1$ and $CR2$ respectively. Initially, $CR1$ is de-energized and $CR2$ is energized. When $IN1$ is turned on and off three times so that three input pulses are counted, the counter turns on and its contacts energize *CR1* and de-energize *CR2*. The counter is reset to the initial state by pulsing its reset coil via *IN2*. The timing diagram for this counter is illustrated in Fig. 5.11.

The preset value of a typical counter can be set in the range 1–999. If a greater count is required counters have to be cascaded. It is also possible to cascade counters with timers.

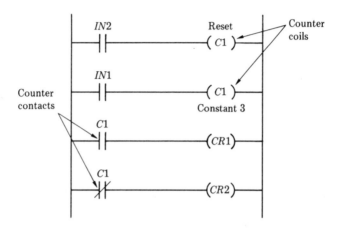

Figure 5.10   Counter circuit.

5.7   Drum sequencer

In ladder, a drum sequencer is a type of counter used for step sequencing. Like any counter it has a count input and a reset input. In addition an up/down counter selection input is included. A table of preset values is set up in memory with each preset value matched with an auxiliary relay. An auxiliary relay is energized when the number of pulses counted by the counter reaches the relay's corresponding preset value.

5.8   Shift register

A number of auxiliary relays, i.e. memory elements, can be grouped together to form a register. For example, a four-bit register could consist of auxiliary relays *AR0*, *AR1*, *AR2* and *AR3* and would be capable of storing a nibble. As explained in Chapter 2, a register in which it is possible to shift stored bits is called a shift register.

Shift registers require at least three inputs, one to load data into the first element of the register, one as the shift command and one for resetting purposes. Consider the four-bit ladder shift register consist-

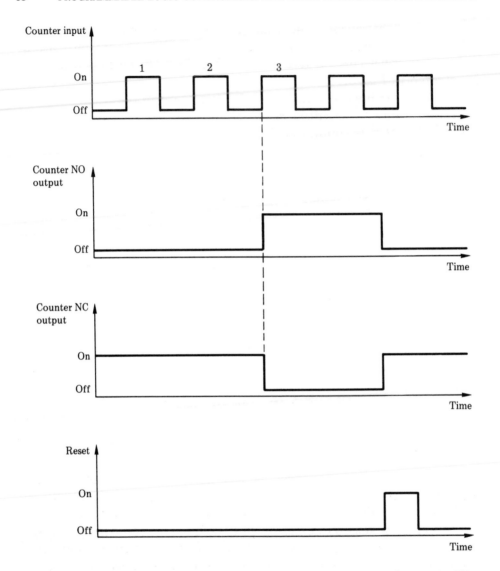

Figure 5.11   Timing diagram for the counter circuit shown in Fig. 5.10

ing of *AR*0, *AR*1, *AR*2 and *AR*3 shown in Fig. 5.12. When contact *IN*1 is turned on a logic one is loaded into the first element of the shift register *AR*0. Contact *IN*2 is connected to the shift right command and when turned on shifts each stored bit to the next right-hand memory location of the register (the last bit is lost). Contact *IN*3 allows the shift register to be cleared so that it stores a series of logic zeros, i.e. off states.

Each memory element or auxiliary relay of a shift register may be used as a contact in a ladder circuit. In Fig. 5.12, *AR*0, *AR*1, *AR*2 and *AR*3 are connected to *CR*1, *CR*2, *CR*3 and *CR*4 respectively. If a logic

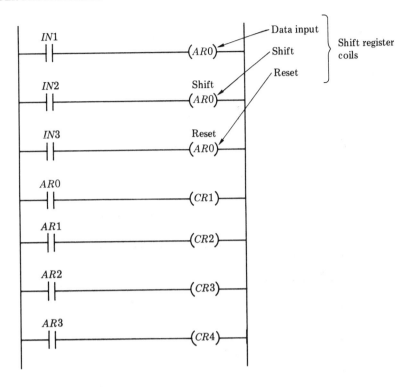

Figure 5.12 Shift register circuit.

one is shifted along the register, the control relays will be energized one after the other as shown by the timing diagram of Fig. 5.13.

5.9 Master control relay

The master control relay (MCR) is a special type of over-ride which is used to de-energize coils in regions of a program. An example of how an MCR may be used is shown in Fig. 5.14. When contact $IN1$ is turned on the MCR de-energizes coils $CR1$ and $CR2$ regardless of the logic states of contacts $IN2$ and $IN3$. The MCR acts only over the region indicated in the figure.

5.10 Jumps

It is often required to jump over part of a ladder circuit when a contact changes its logic state. To skip ladder rungs a jump function is used. Consider Fig. 5.15. When contact $IN1$ is turned on the scan does not execute the part of the program between the jump and jump end function coils.

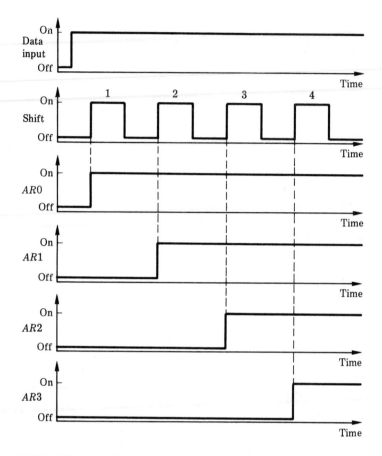

Figure 5.13   Timing diagram for the shift register circuit shown in Fig. 5.12.

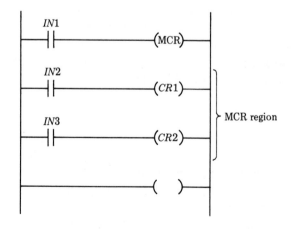

Figure 5.14   Master control relay (MCR).

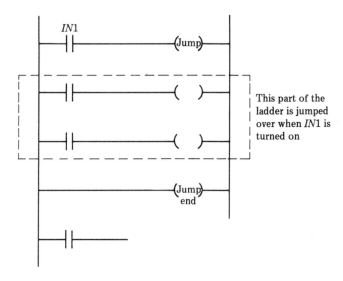

Figure 5.15 Jump function.

5.11 Subroutines

A subroutine is a complete ladder circuit that can be part of another ladder circuit. Subroutine ladders are called from the main ladder as and when required. They must always end with a return instruction (abbreviated to RET).

An example of a subroutine ladder is shown in Fig. 5.16. When

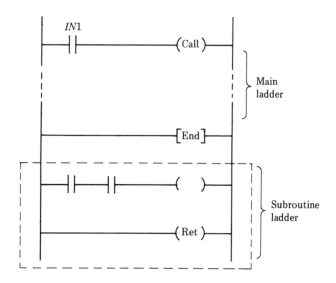

Figure 5.16 Subroutine circuit.

contact *IN*1 in the main part of the ladder is turned on, a call function causes the CPU to start scanning the subroutine ladder and perform the logical operations contained within it. The CPU resumes scanning the main ladder from where it left it when the scan reaches the subroutine's return instruction. Note that the subroutine is scanned only when *IN*1 is turned on.

5.12 Arithmetic functions

Most PLCs are able to do simple arithmetic operations with data stored in registers. Usually three registers are involved, two for storing the data to be manipulated and one for storing the result. With most PLCs the data stored in registers is in BCD form and arithmetic functions perform BCD addition, subtraction, multiplication and division.

Functional instructions allow data to be loaded into a register or data to be read from input contacts into a register so that it may be manipulated by an arithmetic function. If a binary number is stored in a register, functions will exist for converting it into a BCD number. Conversely, it will be possible to convert a BCD number into a binary number.

5.13 I/O functions

A PLC with analogue, high-speed counter and other types of non-standard port will have functions for handling these. Programming normally requires the use of data registers. For example, the operation of producing an analogue output will involve transferring digital data stored in a data register to the input of a DAC. Figure 5.17 shows how this may be represented on a ladder diagram. When *IN*1 is turned on, the 8-bit data register labelled *DR*1 is loaded with the constant 255. A second function coil ensures that this value is output to the PLC's DAC (assumed 8-bit) to produce the full-scale analogue output.

The way in which I/O functions are used depends on the PLC. Individual programming manuals will need to be consulted.

5.14 Alternative programming methods

5.14.1 LOGIC GATES

PLC programs can be developed using the functional symbols for logic gates as shown in Fig. 1.5. The control linkage is established in terms of a circuit consisting of logic gates. The resulting circuit is translated into Boolean code.

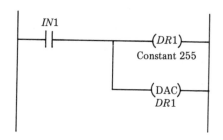

Figure 5.17 DAC function.

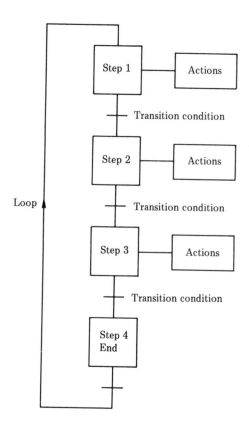

Figure 5.18 GRAFCET method.

5.14.2 GRAFCET

GRAFCET is a step-sequencing programming method for PLCs. A control problem is represented as a sequence of steps and transitions. The controller moves through a GRAFCET program step by step as each transition is cleared. Each step has an associated control action. A transition point represents the closing of a contact.

The GRAFCET programming format, referred to as a state-transition diagram, is illustrated in Fig. 5.18. Being flowchart based, GRAFCET is a half-way house between ladder- and computer-orientated programming languages.

Questions

1. Draw a ladder latch circuit and explain its operation.
2. Draw and label a timing diagram which shows how a delay-on timer operates.
3. Draw and label a timing diagram which shows how a counter having a preset value of five operates.
4. Figure 5.Q4 shows the relationship between the data input and shift-right command of a 4-bit shift register. Show the timing relationship between these signals and the output waveforms generated from memory elements of the register.

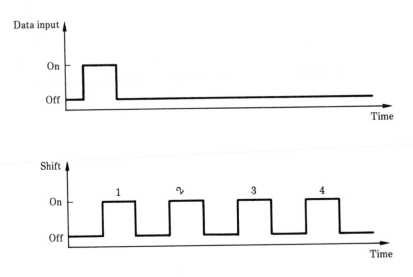

Figure 5.Q4   Timing diagram.

6
Ladder-programming examples

6.1 Introduction

This chapter consists of a series of ladder-programming examples which should allow the development of programming skills. Each example is described in terms of a ladder diagram and its equivalent ladder code. As explained in Chapter 1, ladder code is entered into a programming panel in an address-instruction-data format. We shall use address zero as our starting address.

The general instructions described in Table 6.1 are used for coding the ladder diagrams. Individual instructions are explained in the example programs below. These instructions are common to almost any program and are used by all PLC manufacturers, although notation does vary. Appendix 3 compares the mnemonics used by Mitsubishi and

Table 6.1 Programming instructions

Instruction	Description
LOAD	Load contact
AND	Logical AND operation
OR	Logical OR operation
NOT	Inversion
LOAD NOT	Load inverse
AND NOT	Logical AND NOT operation
OR NOT	Logical OR NOT operation
AND BLOCK	Logically AND two subcircuits
OR BLOCK	Logically OR two subcircuits
RESET	Reset shift register/counter
SHIFT	Shift 1-bit
CONSTANT	Insert constant
END	End ladder

Omron (two leading manufacturers of PLCs) with the general instructions used in this book.

For data, we shall use the notation introduced in Chapters 1 and 5, namely:

IN = input contact
CR = control relay
AR = auxiliary relay
C = counter coil
T = timer coil

When using the ladder programs of this chapter remember to substitute not only the instruction but the data appropriate to your PLC.

Nowadays, many PLCs use a personal computer as a programming panel. A software package is supplied which allows you to develop a ladder circuit on the computer's screen and down load it to the PLC. In this case, there is no need to work out the ladder code.

Finally, all programs assume that the switches connected to the PLC's contacts are normally open.

6.2   Example 1 — Toggling outputs

The circuit shown in Fig. 6.1 shows how to toggle the two control relays $CR1$ and $CR2$ using the contact $IN1$. When $IN1$ is turned off $CR1$ is de-energized and $CR2$ is energized. When $IN1$ is turned on $CR1$ is energized and $CR2$ is de-energized.

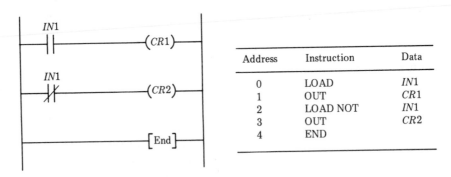

Address	Instruction	Data
0	LOAD	$IN1$
1	OUT	$CR1$
2	LOAD NOT	$IN1$
3	OUT	$CR2$
4	END	

Figure 6.1   Toggling outputs.

6.3   Example 2 — Multiple AND operation

The circuit shown in Fig. 6.2 shows how to logically AND the four contacts $IN1$– $IN4$ and logically OR these with $IN5$ to energize $CR1$. The ladder coding requires the three AND instructions be used.

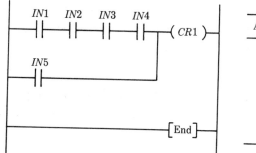

Address	Instruction	Data
0	LOAD	IN1
1	AND	IN2
2	AND	IN3
3	AND	IN4
4	OR	IN5
5	OUT	CR1
6	END	

Figure 6.2 Multiple AND operation.

6.4 Example 3 — Multiple OR operation

The circuit shown in Fig. 6.3 shows how to logically OR the four contacts IN1–IN4 and logically AND these with IN5 to energize CR1. The ladder coding requires that three OR instructions be used.

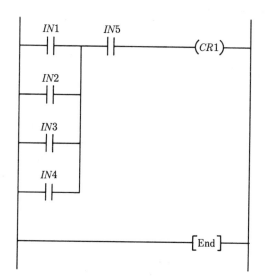

Address	Instruction	Data
0	LOAD	IN1
1	OR	IN2
2	OR	IN3
3	OR	IN4
4	AND	IN5
5	OUT	CR1
6	END	

Figure 6.3 Multiple OR operation.

6.5 Example 4 — Subcircuits

Many complex ladder circuits have to be broken down into subcircuits before they can be converted into ladder code. Each subcircuit is coded as a separate entity. Two subcircuits are combined to form a block. An AND BLOCK instruction is used to connect two subcircuits that are in series. An OR BLOCK instruction is used to connect two subcircuits that are in parallel. AND BLOCK and OR BLOCK instructions may be used more than once on the same ladder line.

Figure 6.4 shows an example of a circuit that must be coded using an AND BLOCK instruction. The circuit is split into the two subcircuits shown. Each subcircuit is converted into ladder code. The AND BLOCK instruction is used to combine the two subcircuits as these are in series with each other.

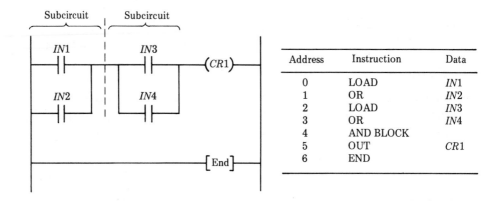

Address	Instruction	Data
0	LOAD	IN1
1	OR	IN2
2	LOAD	IN3
3	OR	IN4
4	AND BLOCK	
5	OUT	CR1
6	END	

Figure 6.4   AND block.

Figure 6.5 shows an example of a circuit that must be coded using an OR BLOCK instruction. This circuit is split into the two subcircuits shown. Each subcircuit is converted into ladder code. The OR BLOCK instruction is used to combine the two subcircuits as these are in parallel with each other.

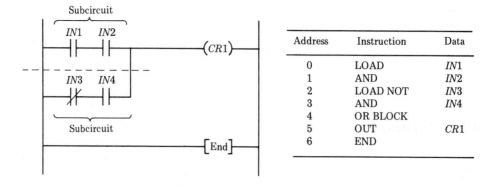

Address	Instruction	Data
0	LOAD	IN1
1	AND	IN2
2	LOAD NOT	IN3
3	AND	IN4
4	OR BLOCK	
5	OUT	CR1
6	END	

Figure 6.5   OR block.

Figures 6.6 and 6.7 show circuits where the ladder code involves using more than one block instruction. In both cases the coding method requires that the block instruction is used immediately after coding the subcircuits.

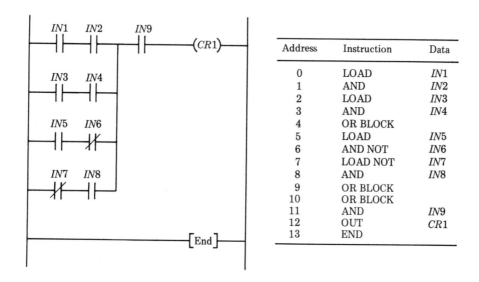

Address	Instruction	Data
0	LOAD	IN1
1	AND	IN2
2	LOAD	IN3
3	AND	IN4
4	OR BLOCK	
5	LOAD	IN5
6	AND NOT	IN6
7	LOAD NOT	IN7
8	AND	IN8
9	OR BLOCK	
10	OR BLOCK	
11	AND	IN9
12	OUT	CR1
13	END	

Figure 6.6   Multiple OR block operation.

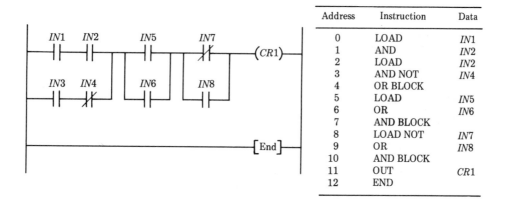

Address	Instruction	Data
0	LOAD	IN1
1	AND	IN2
2	LOAD	IN2
3	AND NOT	IN4
4	OR BLOCK	
5	LOAD	IN5
6	OR	IN6
7	AND BLOCK	
8	LOAD NOT	IN7
9	OR	IN8
10	AND BLOCK	
11	OUT	CR1
12	END	

Figure 6.7   Multiple block operation.

6.6 Example — NAND gate ladder

In question 6 of Chapter 1 you were asked to develop a ladder diagram which used contacts *IN*1 and *IN*2 to produce a NAND gate output on control relay *CR*2. Figure 6.8 shows two ladder circuits which do this. In Fig. 6.8(a) a second ladder line is used to invert the result of logically ANDing *IN*1 with *IN*2. Fig 6.8(b) makes use of De Morgan's theorem $\overline{A.B} = \overline{A} + \overline{B}$.

Note that the circuit of Fig. 6.8(b) uses one ladder line while the circuit shown in Fig. 6.8(a) uses two. Using a single ladder line reduces the coding and also avoids response time problems.

6.7 Example 6 — NOR gate ladder

In question 7 of Chapter 1 you were asked to develop a ladder diagram which used contacts *IN*1 and *IN*2 to produce a NOR gate output on control relay *CR*2. Figure 6.9 shows two ladder circuits which do this.

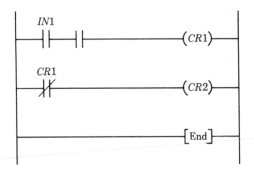

Address	Instruction	Data
0	LOAD	*IN*1
1	AND	*IN*2
2	OUT	*CR*1
3	LOAD NOT	*CR*1
4	OUT	*CR*2
5	END	

(a)

Address	Instruction	Data
0	LOAD NOT	*IN*1
1	OR NOT	*IN*2
2	OUT	*CR*2
3	END	

(b)

Figure 6.8 NAND gate ladder circuits.

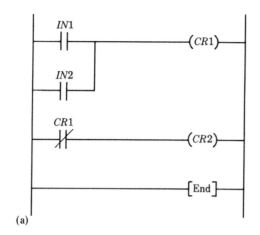

Address	Instruction	Data
0	LOAD	IN1
1	OR	IN2
2	OUT	CR1
3	LOAD NOT	CR1
4	OUT	CR2
5	END	

(a)

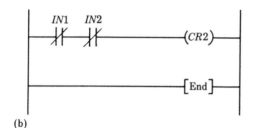

Address	Instruction	Data
0	LOAD NOT	IN1
1	AND NOT	IN2
2	OUT	CR2
3	END	

(b)

Figure 6.9   NOR gate ladder circuits.

In Fig. 6.9(a) a second ladder line is used to invert the result of logically ORing *IN*1 with *IN*2. Figure 6.9(b) makes use of De Morgan's theorem $\overline{A+B} = \overline{A}.\overline{B}$, which reduces the coding.

6.8   Example 7 — XOR gate ladder

In Question 8 of Chapter 1 you were asked to develop a ladder diagram which used contacts *IN*1 and *IN*2 to produce an XOR gate output on control relay *CR*2. Figure 6.10 shows three ladder circuits which do this. The ladder circuit of Fig. 6.10(a) is basically an OR circuit for *IN*1 and *IN*2 with *CR*1 added to prevent *CR*2 being energized when *IN*1 and *IN*2 are turned on together. Likewise, both the circuits shown in Figures 6.10(b) and (c) are basically OR circuits for *IN*1 and *IN*2 but use normally closed contacts for *IN*1 and *IN*2 to prevent *CR*2 being energized when *IN*1 and *IN*2 are turned on together. The ladder circuit of 6.10(b) is coded by dividing it up into two subcircuits which can be combined using an AND BLOCK instruction. The ladder circuit of Fig. 6.10(c) is coded by dividing it up into two subcircuits which can be combined using an OR BLOCK instruction.

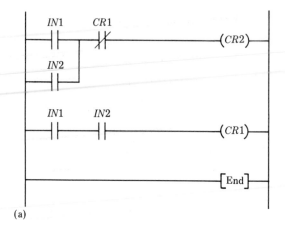

Address	Instruction	Data
0	LOAD	IN1
1	OR	IN2
2	AND NOT	CR1
3	OUT	CR2
4	LOAD	IN1
5	AND	IN2
6	OUT	CR1
7	END	

(a)

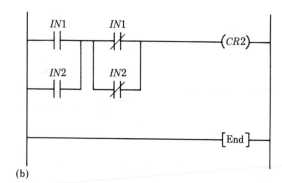

Address	Instruction	Data
0	LOAD	IN1
1	OR	IN2
2	LOAD NOT	IN1
3	OR NOT	IN2
4	AND BLOCK	
5	OUT	CR2
6	END	

(b)

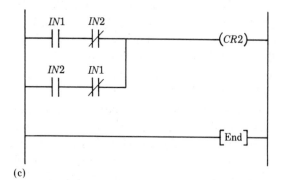

Address	Instruction	Data
0	LOAD	IN1
1	AND NOT	IN2
2	LOAD	IN2
3	AND NOT	IN1
4	OR BLOCK	
5	OUT	CR2
6	END	

(c)

Figure 6.10 XOR gate ladder circuits.

6.9 Example 8 — Resetting a latch with an auxiliary relay

The circuit shown in Fig. 6.11 shows how to use an auxiliary relay to reset a latch circuit. When contact *IN1* (assumed to be a pushbutton contact) is momentarily pressed *CR1* is energized. *CR1* is latched to keep itself energized when the pushbutton is released. When contacts *IN2* and *IN3* are turned on together, the auxiliary relay *AR1* is energized and breaks the latch circuit with the result that *CR1* is de-energized. Thus, the use of an auxiliary relay allows a logical operation of contacts to reset a latch.

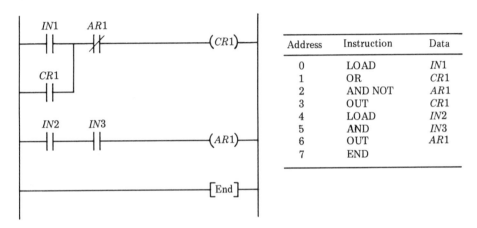

Address	Instruction	Data
0	LOAD	*IN1*
1	OR	*CR1*
2	AND NOT	*AR1*
3	OUT	*CR1*
4	LOAD	*IN2*
5	AND	*IN3*
6	OUT	*AR1*
7	END	

Figure 6.11   Resetting a latch with an auxiliary relay.

6.10 Example 9 — Resetting a latch with a counter

The circuit shown in Fig. 6.12 shows how a counter may be used to reset a latch circuit. When contact *IN1* (assumed to be a pushbutton contact) is momentarily pressed *CR1* is energized and counter *C1* is reset. *CR1* is latched to keep itself energized when the pushbutton is released. When *IN2* is turned on and off three times the counter contact breaks the latch circuit and *CR1* is de-energized.

6.11 Example 10 — Resetting a timer

The circuit shown in Fig. 6.13 shows a timer circuit which may be reset. When contact *IN1* (assumed to be a pushbutton contact) is momentarily pressed the 5 s delay-on type starts measuring time. The auxiliary relay *AR1* is used to keep the timer energized when the pushbutton is released. After five seconds the timer energizes *CR1* unless *IN2* is turned on. When *IN2* is turned on the latch circuit is broken and the timer is reset.

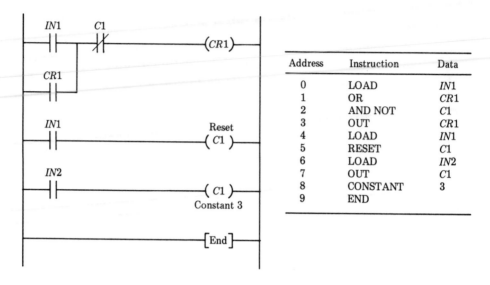

Address	Instruction	Data
0	LOAD	IN1
1	OR	CR1
2	AND NOT	C1
3	OUT	CR1
4	LOAD	IN1
5	RESET	C1
6	LOAD	IN2
7	OUT	C1
8	CONSTANT	3
9	END	

Figure 6.12   Resetting a latch with a counter.

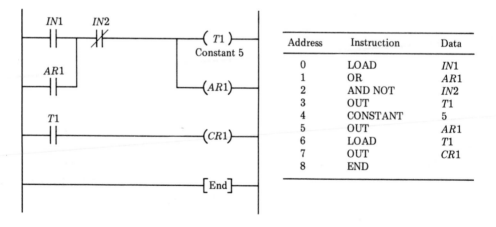

Address	Instruction	Data
0	LOAD	IN1
1	OR	AR1
2	AND NOT	IN2
3	OUT	T1
4	CONSTANT	5
5	OUT	AR1
6	LOAD	T1
7	OUT	CR1
8	END	

Figure 6.13   Timer circuit with reset.

In this circuit, we are in effect giving $IN2$ 5 s to reset the timer $T1$ before $CR1$ is energized. This 'time out' programming technique can be very useful in certain situations.

6.12 Example 11 — Cyclic timer

A circuit which repeatedly turns a control relay (or other type of coil) on and off at regular intervals is called a cyclic timer. A cyclic timer circuit is shown in Fig. 6.1. While $IN1$ is turned on, the delay-on timers $T1$ and

$T2$ set and reset each other with the result that $CR1$ is clocked on and off at 5 s intervals.

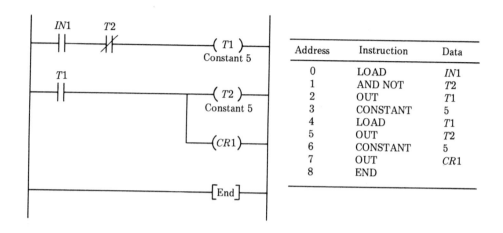

Address	Instruction	Data
0	LOAD	$IN1$
1	AND NOT	$T2$
2	OUT	$T1$
3	CONSTANT	5
4	LOAD	$T1$
5	OUT	$T2$
6	CONSTANT	5
7	OUT	$CR1$
8	END	

Figure 6.14   Cyclic timer.

6.13   Example 12 — Delay-off timer

Many PLCs provide the programmer with delay-on type timers only. A delay-off timer must be created by using a delay-on timer to reset a latch circuit. Consider the circuit shown in Fig. 6.15. When contact $IN1$ (assumed to be a pushbutton contact) is momentarily pressed $CR1$ is energized and the 5 s delay-on timer $T1$ starts measuring time. $CR1$ is latched to keep the timer and itself energized when the pushbutton is released. After 5 s, the timer contact is activated and breaks the latch circuit so that $CR1$ is de-energized. Thus, when $IN1$ is momentarily pressed the circuit energizes $CR1$ for 5 s which is the action of a delay-off timer.

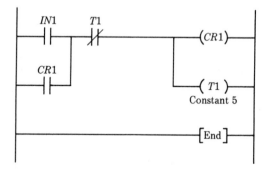

Address	Instruction	Data
0	LOAD	$IN1$
1	OR	$CR1$
2	AND NOT	$T1$
3	OUT	$CR1$
4	OUT	$T1$
5	CONSTANT	5
6	END	

Figure 6.15   Delay-off timer.

6.14 Example 13 — Sequencing outputs using delay-off timers

The circuit shown in Fig. 6.16 shows how to use three delay-off timer circuits to repeatedly energize and de-energize the control relays $CR1$,

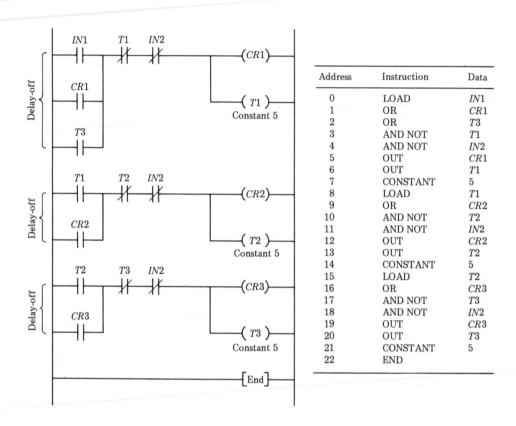

Figure 6.16   Sequencing using delay-off timers.

$CR2$ and $CR3$ one after the other as indicated by the timing diagram of Fig. 6.17. The sequencing is started when $IN1$ (assumed to be a pushbutton) is momentarily pressed, as this initiates the first delay-off timer circuit which keeps $CR1$ energized for 5 s. When $CR1$ is de-energized the second delay-off circuit is initiated via $T1$ and keeps $CR2$ energized for 5 s. When $CR2$ is de-energized the third delay-off circuit is initiated via $T2$ and keeps $CR3$ energized for 5 s. When $CR3$ is de-energized the first delay-off circuit is again initiated via $T3$, so that the sequencing is repeated until contact $IN2$ (assumed to be a pushbutton) is momentarily pressed. When pressed, $IN2$ breaks all of the latch circuits and so stops the sequencing action.

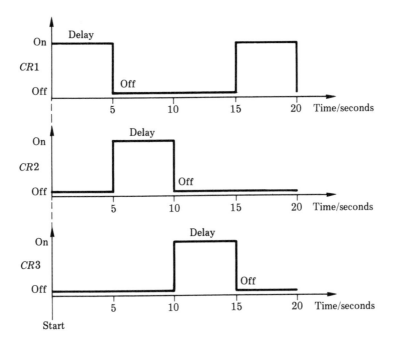

Figure 6.17   Timing diagram for the sequencing circuit shown in Fig. 6.16.

6.15   Example 14 — Sequencing outputs when overlap is required

The circuit shown in Fig. 6.18 is an adapted version of that in Fig. 6.16. It allows a 1 s overlap between the control relays turning on and off as shown by the timing diagram of Fig. 6.19. The overlap is achieved by inserting timers to start the next delay-off circuit 1 s before the previous delay-off action finishes.

6.16   Example 15 — Sequencing using a shift register

The circuit shown in Fig. 6.20 shows how to use a 4-bit shift register to energize control relays $CR1$, $CR2$, $CR3$ and $CR4$ one after the other. Contact $IN1$ is latched to provide an input to the first element of the shift register. Contact $IN2$ is the shift input and is also used to break the data input latch circuit. When $IN1$ is momentarily turned on and $IN2$ turned on and off four times the control relays $CR1$ to $CR4$ are energized one after the other.

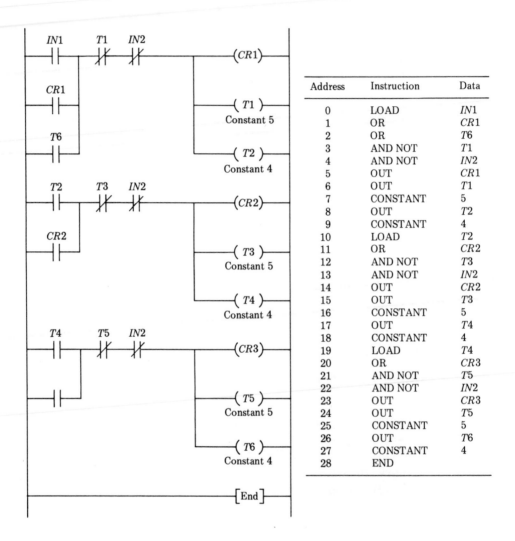

Address	Instruction	Data
0	LOAD	IN1
1	OR	CR1
2	OR	T6
3	AND NOT	T1
4	AND NOT	IN2
5	OUT	CR1
6	OUT	T1
7	CONSTANT	5
8	OUT	T2
9	CONSTANT	4
10	LOAD	T2
11	OR	CR2
12	AND NOT	T3
13	AND NOT	IN2
14	OUT	CR2
15	OUT	T3
16	CONSTANT	5
17	OUT	T4
18	CONSTANT	4
19	LOAD	T4
20	OR	CR3
21	AND NOT	T5
22	AND NOT	IN2
23	OUT	CR3
24	OUT	T5
25	CONSTANT	5
26	OUT	T6
27	CONSTANT	4
28	END	

Figure 6.18   Sequencing using overlapping delay-off timers.

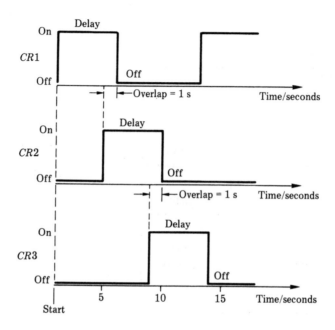

Figure 6.19 Timing diagram for the sequencing circuit shown in Fig. 6.18.

The circuit shown in Fig. 6.21 is an adapted version of Fig. 6.20. It shows how a cyclic timer circuit may be used with a shift register circuit to repeatedly energize and de-energize the control relays $CR1$, $CR2$, $CR3$, $CR4$ one after the other. A 5-bit shift register is used. The fifth bit ensures that the sequencing is repeated.

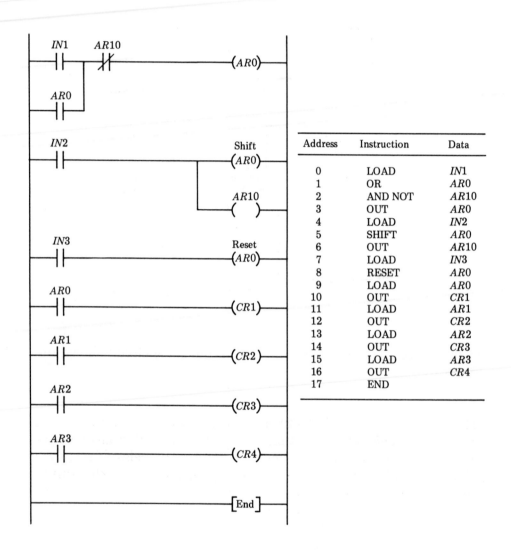

Address	Instruction	Data
0	LOAD	IN1
1	OR	AR0
2	AND NOT	AR10
3	OUT	AR0
4	LOAD	IN2
5	SHIFT	AR0
6	OUT	AR10
7	LOAD	IN3
8	RESET	AR0
9	LOAD	AR0
10	OUT	CR1
11	LOAD	AR1
12	OUT	CR2
13	LOAD	AR2
14	OUT	CR3
15	LOAD	AR3
16	OUT	CR4
17	END	

Figure 6.20   Sequencing using a shift register.

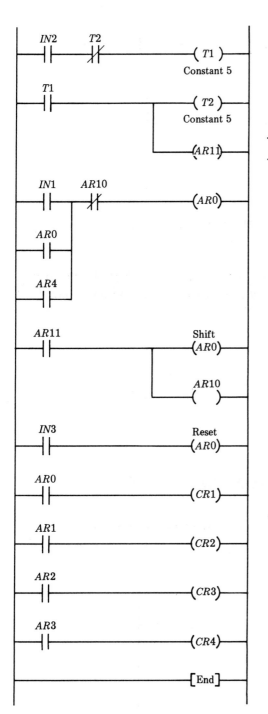

Address	Instruction	Data
0	LOAD	IN2
1	AND NOT	T2
2	OUT	T1
3	CONSTANT	5
4	LOAD	T1
5	OUT	T2
6	CONSTANT	5
7	OUT	AR11
8	LOAD	IN1
9	OR	AR0
10	OR	AR4
11	AND NOT	AR10
12	OUT	AR0
13	LOAD	AR11
14	SHIFT	AR0
15	OUT	AR10
16	LOAD	IN3
17	RESET	AR0
18	LOAD	AR0
19	OUT	CR1
20	LOAD	AR1
21	OUT	CR2
22	LOAD	AR2
23	OUT	CR3
24	LOAD	AR3
25	OUT	CR4
26	END	

Figure 6.21 Cyclic shift register.

Questions

1. Convert the ladder diagram in Fig. 6.Q1 into ladder code.

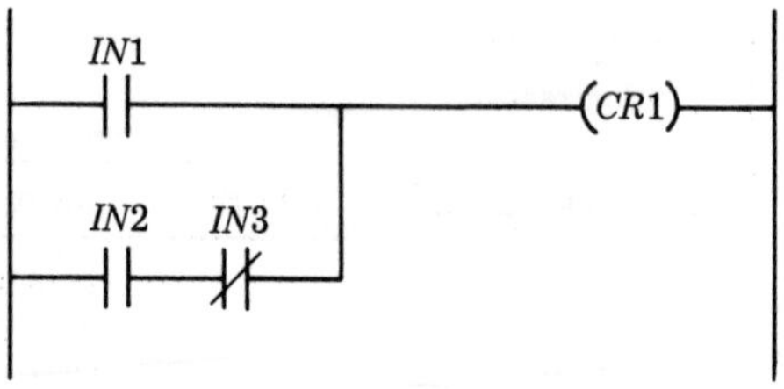

Figure 6.Q1   Circuit.

2. The circuit of question one may be rearranged to that shown in Fig. 6.Q2. Write down the ladder code for this circuit. Which is the best circuit to use and why?

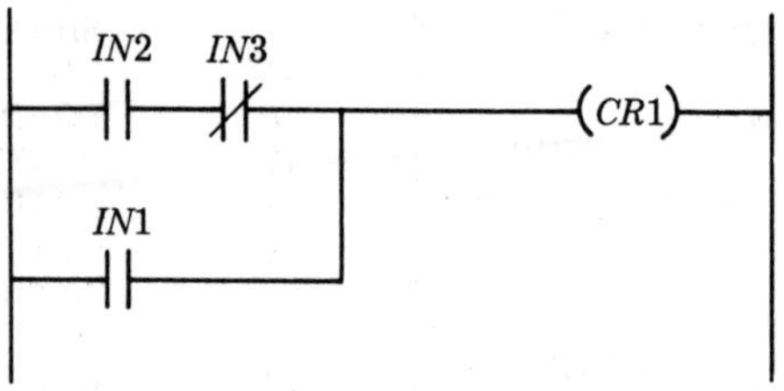

Figure 6.Q2   Circuit.

3. Convert the ladder diagram in Fig. 6.Q3 into ladder code.

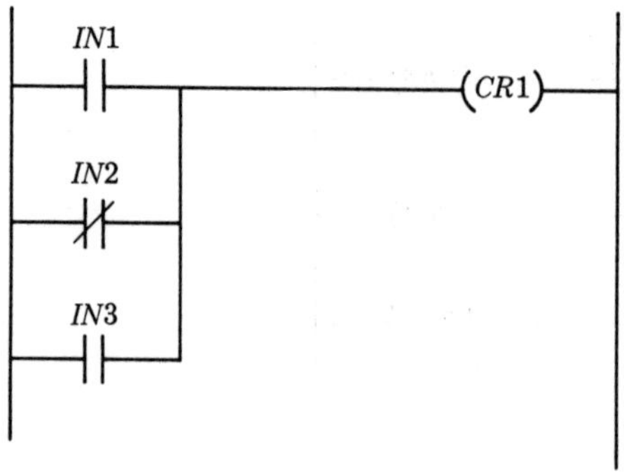

Figure 6.Q3   Circuit.

4. By dividing the circuit shown in Fig. 6.Q4 into subcircuits, convert it into ladder code.

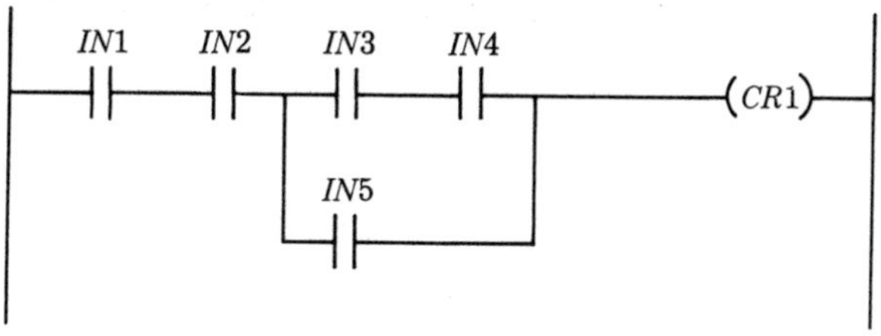

Figure 6.Q4 Circuit.

5. By dividing the circuit shown in Fig. 6.Q5 into subcircuits convert it into ladder code.

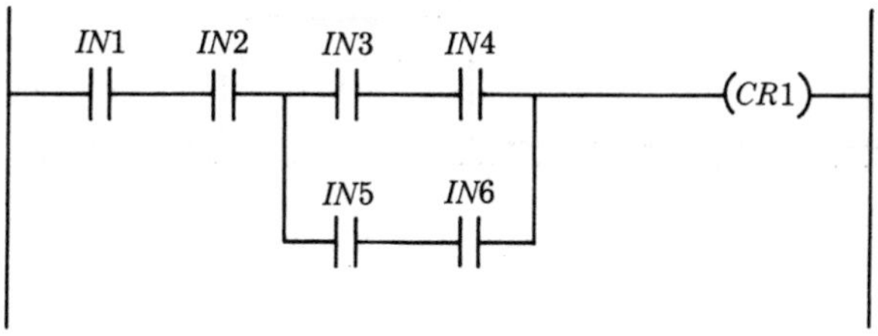

Figure 6.Q5 Circuit.

6. The circuit in Fig. 6.Q6 uses contacts IN1, IN2 and IN3 to energize the three control relays one after the other. Convert this circuit into ladder code.

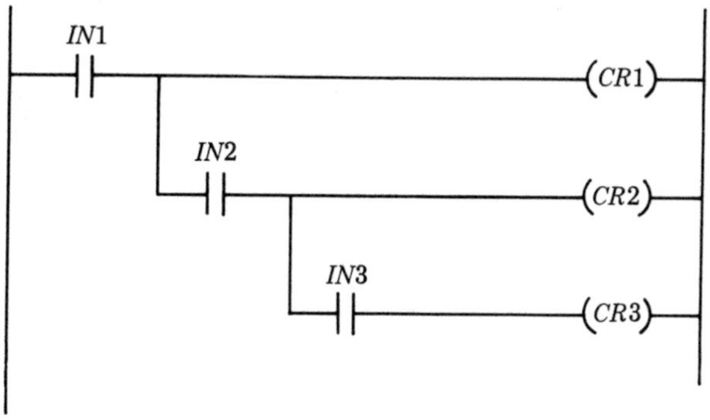

Figure 6.Q6 Circuit.

7. Develop and code a ladder circuit which produces the timing sequence illustrated in Fig. 6.Q7 for the three control relays $CR1$, $CR2$ and $CR3$. The sequencing is to be started by the momentary contact of $IN1$ and reset by the momentary contact of $IN2$.

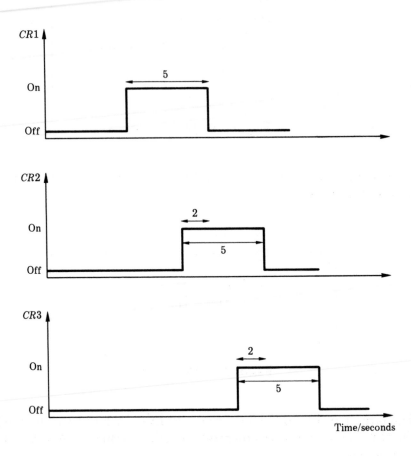

Figure 6.Q7   Timing diagram.

7
Application examples

7.1 Introduction

This chapter is concerned with how PLCs are used in typical control applications. As the successful solution of a control problem requires an understanding of the system being controlled, each example in this chapter will be divided as follows:

1. The control problem.
2. Description of control elements.
3. Programming solution.

The programming solutions are expressed as ladder and BASIC programs. The analogy between BASIC and ladder programming is aimed at helping the programmer with a background in BASIC make the transition to ladder. However, it should be pointed out that BASIC is geared to reading/writing data bytes to and from 8-bit parallel computer-type input/output ports.

If you are a BASIC programmer you will know what an 8-bit parallel port is. If not, Fig. 7.1 illustrates an 8-bit parallel port used to input data. Eight switches labelled SW0–SW7 are connected to the port's data bits 0–7. The user determines which of the eight switches are turned on from the data value (byte) read from the port. This data value ranges from 0 (all switches off) to 255 (all switches on). For example, when switch SW0 is turned on with all other switches turned off the data value read from the parallel port is 1. Likewise, when SW7 is turned on with all other switches turned off the data value read from the parallel port is 128. With SW0 and SW7 turned on together with all other switches turned off the data value read from the parallel port is 129 (i.e. 128 + 1). Conversely, data in the range 0 to 255 may be transmitted to an output port to set and reset individual port lines.

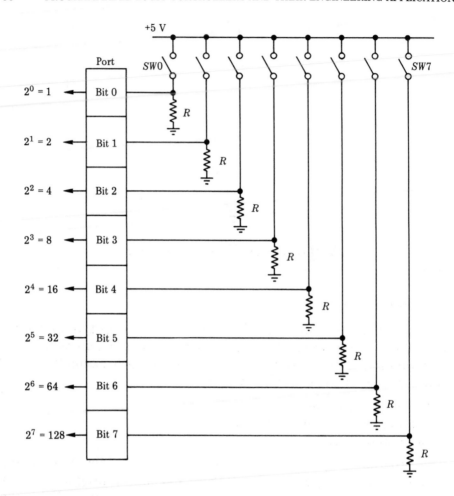

Figure 7.1 Eight-bit parallel computer port.

Finally, in all examples, it is assumed that the base unit of the controller (regardless of its type) has opto-isolated input ports which require sourcing and relay driven output ports.

7.2 Example 1 — Moving a pneumatic piston

7.2.1 THE CONTROL PROBLEM

The task of the PLC is to move the piston in and out of the pneumatic cylinder shown in the arrangement of Fig. 7.2. When switch $SW1$ is momentarily turned on piston A is to move out of the cylinder (referred to as the $A+$ direction). When switch $SW2$ is momentarily turned on piston A is to move into the cylinder (referred to as the $A-$ direction).

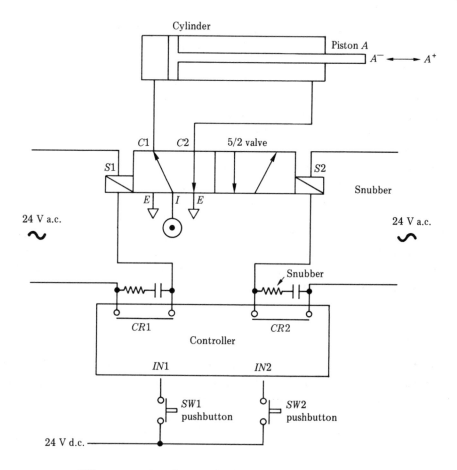

Figure 7.2 Control of a pneumatic piston.

7.2.2 DESCRIPTION OF CONTROL ELEMENTS

The pneumatic cylinder comprises a piston which is driven by pressurized air. The air flow direction to the cylinder is controlled by the solenoid-driven 5/2 directional control valve (see below for the description of a 5/2 valve). The arrangement is such that when solenoid $S1$ of the valve is energized (with solenoid $S2$ de-energized) air enters through $C1$ and exhausts through $C2$ causing the piston to move out of the cylinder ($A+$). When solenoid $S2$ is energized (with solenoid $S1$ de-energized) air enters through $C2$ and exhausts through $C1$, causing, the piston to move into the cylinder (A^-).

For readers who are unfamiliar with pneumatics, the 5/2 valve is a standard directional control valve. It is able to change the direction of air flow by the movement of an internal component called the spool. In pneumatics special symbols are used to describe valves. These are drawn as a series of boxes containing arrows which indicate the air flow direction of the inlets and outlets when the spool is in a particular position. A 5/2 valve is called 5/2 because it has a total of five inlets and outlets and two spool positions (i.e. its symbol has two boxes). The valve solenoids are used to move the spool to either position. If the pneumatic arrangement is to be built the two outlets labelled E (for exhaust) can be left unconnected as this allows air to be exhausted to the atmosphere. Inlet I is connected to the pressurized air supply.

Interfacing the 24 V a.c. solenoids to a controller with relay output ports is straightforward as shown in Fig. 7.2. Snubbers should be connected across $CR1$ and $CR2$ respectively. The switches can be activated either by the user or by the action of the piston itself. In the latter case, either the switches can be mounted externally so that they are activated by the piston at either end of its stroke or, provided the plunger is magnetic, Reed switches mounted at either end of the cylinder may be used (see Chapter 3).

7.2.3   PROGRAMMING

A ladder program for the control action required is shown in Fig. 7.3. Two latch circuits are used. When contact $IN1$ is momentarily turned on $CR1$ is energized via the latch circuit. When $IN2$ is momentarily turned on $CR1$ is de-energized ($IN2$ resets its latch) and $CR2$ energized via a second latch circuit. The second latch circuit is reset when $IN1$ is turned on.

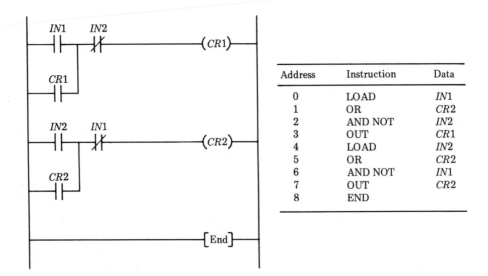

Address	Instruction	Data
0	LOAD	$IN1$
1	OR	$CR2$
2	AND NOT	$IN2$
3	OUT	$CR1$
4	LOAD	$IN2$
5	OR	$CR2$
6	AND NOT	$IN1$
7	OUT	$CR2$
8	END	

Figure 7.3   Circuit for controlling the pneumatic piston.

A BASIC program which emulates the control action required is shown below.

```
10 A = INP(port1)
20 IN1 = A AND 1
30 IN2 = A AND 2: IF IN2 = 2 THEN IN2 = 1
40 IF IN1 = 1 AND IN2 = 0 THEN OUT port 2,1
50 IF IN2 = 1 AND IN1 = 0 THEN OUT port 2,2
60 IF IN1 = 1 AND IN2 = 1 THEN OUT port 2,0
70 GOTO 10
```

It assumes (a) that port 1 is an 8-bit input port with switches $SW1$ and $SW2$ connected to bits 0 and 1 respectively and (b) that port 2 is an 8-bit output port with solenoids $S1$ and $S2$ connected to bits 0 and 1 respectively.

Line 10 reads a data byte from the parallel input port. Lines 20 and 30 use logical AND functions to obtain bit values for $SW1$ and $SW2$ respectively. This way of using an AND function to isolate a single bit is called masking. If $SW1$ is on and $SW2$ is off line 40 turns on bit 0 of port 2 (i.e. solenoid $S1$). If $SW2$ is on and $SW1$ is off line 50 turns on bit 1 of port 2 (i.e. solenoid $S2$). Line 60 ensures that if both $SW1$ and $SW2$ are turned on together the outputs are de-energized. Line 70 transfers control back to line 10 and so emulates a PLC's repetitive scanning action.

7.3 Example 2 — Cyclic movement of a piston using a timer

7.3.1 THE CONTROL PROBLEM

The task of the PLC is to continuously move, with the use of a timer, a pneumatic piston in and out of its cylinder. The arrangement is the same as that shown in Fig. 7.2 except that switches $SW1$ and $SW2$ are not used.

7.3.2 DESCRIPTION OF CONTROL ELEMENTS

The pneumatic control elements were described in the previous example. Automatic control requires the use of a programmable timer.

7.3.3 PROGRAMMING SOLUTION

A ladder program which implements the control action required is shown in Fig. 7.4. It is based on a cyclic timer circuit discussed in Chapter 6. The two timers $T1$ and $T2$ set and reset each other at 5 s intervals with the result that $CR1$ and $CR2$ are toggled (i.e. switched so that when $CR1$ is energized $CR2$ is de-energized and vice versa) every 5 s.

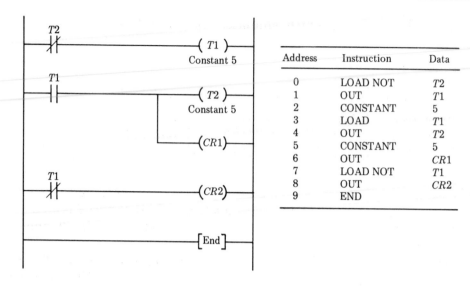

Address	Instruction	Data
0	LOAD NOT	T2
1	OUT	T1
2	CONSTANT	5
3	LOAD	T1
4	OUT	T2
5	CONSTANT	5
6	OUT	CR1
7	LOAD NOT	T1
8	OUT	CR2
9	END	

Figure 7.4 Cyclic timer circuit for controlling the piston.

A BASIC program which implements this control action is as shown below.

```
10 OUT port 2,1
20 GOSUB 60
30 OUT port 2,2
40 GOSUB 60
50 GOTO 10
60 REM Timer subroutine
70 FOR T = 1 to 50000:NEXT T
80 RETURN
```

It assumes that port 2 is an 8-bit output port with solenoid $S1$ connected to bit 0 and solenoid $S2$ connected to bit 1. Line 10 energizes solenoid $S1$ and de-energizes $S2$. Line 20 calls the timer subroutine which provides a fixed delay. This keeps solenoid $S1$ energized and $S2$ de-energized for a fixed period. Line 30 energizes solenoid $S2$ and de-energizes $S1$. Line 40 calls the timer subroutine again and to keep the solenoids in this 'flipped' state for a fixed period. Line 50 transfers control to line 10 and so emulates a PLC's repetitive scanning action.

The timer subroutine uses a FOR loop to cause the fixed delay. Changing the limit specified in the FOR loop changes the delay period and is equivalent to the preset value of a ladder timer. The exact delay period is system dependent, but in this example it is assumed a loop value of 10 000 provides a 1 s delay.

7.4 Example 3 — Automatic sequencing of three pneumatic pistons

7.4.1 THE CONTROL PROBLEM

The task of the PLC is to operate piston A, followed by piston B and finally piston C, using the arrangement shown in Fig. 7.5. This sequence $A+, A-, B+, B-, C+, C-$ is to be repeated while switch $SW1$, connected to input port $IN1$, is turned on.

7.4.2 DESCRIPTION OF CONTROL ELEMENTS

The three pneumatic pistons shown in Fig. 7.5 are controlled from three solenoid-driven 5/2 directional control valves. To accomplish the required control task, the PLC has to energize the valve solenoids in the sequence $S1, S2, S3, S4, S5, S6$. Each solenoid has to be held energized for a short period to allow time for the piston to move. The sequencing of these solenoids is to be repeated while $SW1$ is turned on.

7.4.3 PROGRAM SOLUTION

There are a number of ways in which this action may be implemented in ladder. A series of delay-off timer circuits may be used as shown in Fig. 7.6. Alternatively, a shift register could be used in conjunction with a cyclic timer. Both types of circuit are explained in Chapter 6.

A BASIC program which implements this control action is shown below.

```
10 A = INP (port1)
15 IN1 = A and 1
20 WHILE IN1 > 0
30 FOR I = 0 TO 5
40 OUT port 2,2 ↑ I
50 GOSUB 100
60 NEXT I
70 A = INP (port1)
75 IN1 = A AND 1
80 WEND
90 GOTO 10
100 REM Timer subroutine
110 FOR T = 1 TO 50000:NEXT T
120 RETURN
```

The program assumes that switch $SW1$ is connected to bit 0 of an input port called port 1. Solenoids $S1$ to $S6$ are connected to bits 0–5 of an output port called port 2. A WHILE statement is used to ensure that solenoids are energized only when $SW1$ is turned on. In BASIC the

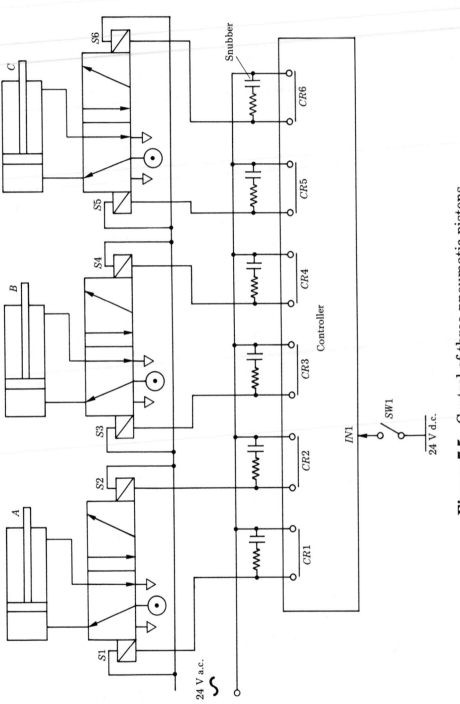

Figure 7.5 Control of three pneumatic pistons.

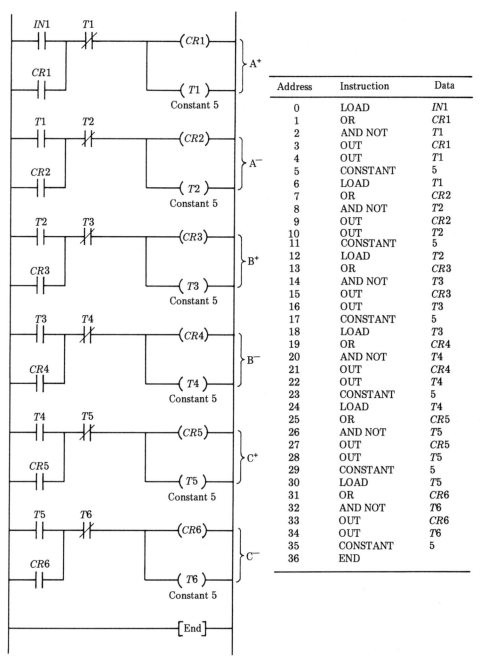

Address	Instruction	Data
0	LOAD	IN1
1	OR	CR1
2	AND NOT	T1
3	OUT	CR1
4	OUT	T1
5	CONSTANT	5
6	LOAD	T1
7	OR	CR2
8	AND NOT	T2
9	OUT	CR2
10	OUT	T2
11	CONSTANT	5
12	LOAD	T2
13	OR	CR3
14	AND NOT	T3
15	OUT	CR3
16	OUT	T3
17	CONSTANT	5
18	LOAD	T3
19	OR	CR4
20	AND NOT	T4
21	OUT	CR4
22	OUT	T4
23	CONSTANT	5
24	LOAD	T4
25	OR	CR5
26	AND NOT	T5
27	OUT	CR5
28	OUT	T5
29	CONSTANT	5
30	LOAD	T5
31	OR	CR6
32	AND NOT	T6
33	OUT	CR6
34	OUT	T6
35	CONSTANT	5
36	END	

Figure 7.6 Circuit for sequencing the pistons.

symbol ↑ is the exponential operator. Thus, the value output to port 2 is the power of two which sets the appropriate solenoid. Individual bits of an 8-bit parallel port are set by the numbers shown in Fig. 7.1.

7.5   Example 4 — Start/stop motor control

7.5.1   THE CONTROL PROBLEM

The task of the PLC is to start a motor when a pushbutton labelled START is pressed and to stop the motor when a pushbutton labelled STOP is pressed. This is a typical problem encountered by production engineers wishing to control a conveyor motor.

7.5.2   DESCRIPTION OF CONTROL ELEMENTS

Figure 7.7 shows the START and STOP pushbuttons connected to $IN1$ and $IN2$ of a PLC. A small d.c. motor is connected to $CR1$. Larger industrial type motors will need to be switched using a secondary device such as a contactor as explained in Chapter 3.

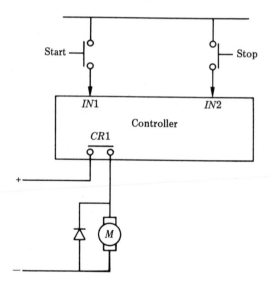

Figure 7.7   Start/stop motor control.

7.5.3   PROGRAM SOLUTION

The simple latch circuit shown in Fig. 7.8 is used to start and stop the motor. When the start pushbutton connected to $IN1$ is pressed the control relay $CR1$ is energized and keeps the motor running once the pushbutton is released. The motor is stopped by pressing the stop pushbutton connected to $IN2$ as this breaks the latch circuit.

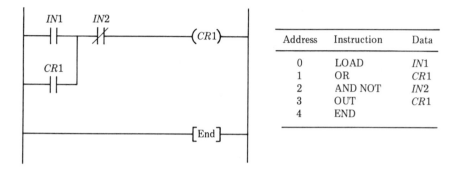

Figure 7.8 Latch circuit for start/stop motor control.

A BASIC program which emulates this control action is shown below.

10 A = INP (port1)
20 IN1 = A AND 1
30 IN2 = A AND 2: IF IN2 = 2 THEN IN2 = 1
40 IF IN1 = 1 AND IN2 = 0 THEN OUT port 2,1
50 IF IN2 = 1 AND IN1 = 0 THEN OUT port 2,0
60 IF IN1 = 1 AND IN2 = 1 THEN OUT port 2,0
70 GOTO 10

It assumes (a) that port 1 is an 8-bit input port with the start and stop pushbuttons connected to bits 0 and 1 respectively, and (b) that port 2 is an 8-bit output port with the motor connected to bit zero.

7.6 Example 5 — Motor speed control

7.6.1 THE CONTROL PROBLEM

The task of the PLC is to ramp a motor to a set speed using the arrangement shown in Fig. 7.9.

7.6.2 DESCRIPTION OF CONTROL ELEMENTS

Invariably in motor speed control problems the PLC has to generate an analogue signal using a DAC. This is because most motor drives require an analogue signal input which proportionally controls motor speed. The speed signal is usually 0–10 V.

The motor arrangement shown in Fig. 7.9 could be either a d.c. motor controlled by a converter drive or an a.c. motor controlled by an inverter drive. The DAC is shown as a separate unit but is often built into the PLC. The analogue signal from the DAC may need to be amplified or impedance matched to the drive (see Section 4.5).

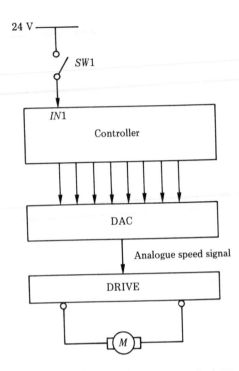

Figure 7.9 Motor speed control.

7.6.3 PROGRAM SOLUTION

The application requires that the speed of a motor is gradually increased by enlarging the analogue value produced by the DAC. An 8-bit DAC accepts numbers in the range 0 to 255 where 0 produces no output and 255 produces full-scale output (assumed to be 10 V). A ladder which ramps the motor to full speed will have the form shown in Fig. 7.10. When $IN1$ is turned off the data register DR stores zero so the DAC has no output. When $IN1$ is turned on the value stored in the data register is incremented by one and consequently the DAC may be ramped. Alternatively, a cyclic timer circuit may be used to increment the ramp data at a regular rate to increase the motor's speed.

A BASIC program which implements this control action is shown below.

```
10 A = INP (port 1)
20 IN1 = A AND 1
30 IF IN1 > 0 THEN GOTO 40 ELSE GOTO 10
40 FOR I = 0 TO 255
50 OUT port 2, I
60 NEXT I
```

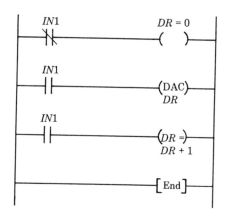

Figure 7.10 Ladder circuit for ramping motor speed.

The program assumes (a) that *SW1* is connected to bit 0 of an input port called port 1 and (b) port 2 is an 8-bit parallel port connected directly to the digital input of the DAC. The program uses a FOR loop instruction to gradually increase the digital input to the DAC and hence the analogue speed command. To ramp the motor back down to zero speed a second FOR loop may be used in which the variable is decremented. For example:

```
70 FOR I = 255 TO 0 STEP −1
80 OUT port 2, I
90 NEXT I
```

A delay subroutine may be used inside the loop. Changing the delay changes the ramp rate.

7.7 Example 6 — On/off temperature control

7.7.1 THE CONTROL PROBLEM

In this application the PLC is to act as an on/off temperature controller using the arrangement shown in Fig. 7.11. An on/off temperature controller turns a heater on when the sensed temperature is below a set point value and off otherwise.

7.7.2 DESCRIPTION OF CONTROL ELEMENTS

Analogue temperature sensors such as thermocouples and diodes can be interfaced to a PLC using a comparator amplifier (see Chapter 3). In this example, it is assumed that an interface shown in Fig. 3.8 has been used for the temperature sensor. When the sensed temperature is below

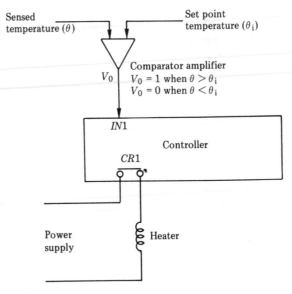

Figure 7.11   On/off temperature control.

the set point value the comparator amplifier produces a logic zero. When the sensed temperature is above the set point value the comparator amplifier produces a logic one.

7.7.3 PROGRAM SOLUTION

The ladder solution to this problem is straightforward. When the comparator amplifier is off the heater output should be on and vice versa. A simple one-line ladder program (see Fig. 7.12) achieves this action.

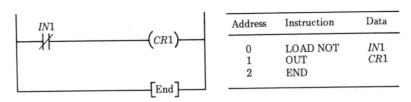

Figure 7.12   Circuit for on/off temperature control.

The BASIC program also has a simple form as shown below:

```
10 A = INP (port 1)
20 IN1 = A AND 1
30 IF IN1 = 0 THEN OUT port 2,1
40 IF IN1 = 1 THEN OUT port 2,0
50 GOTO 10
```

The program assumes (a) that the output of the comparator amplifier is connected to bit 0 of an 8-bit parallel input port called port 1 and (b) that the heater is switched from bit 0 of an 8-bit parallel output port called port 2.

With this type of control action the temperature oscillates about the set point value – a situation which is referred to as hunting (see Fig. 7.13). Oscillations are proportional to the time constant of the system being heated. If the time constant is short, the oscillations are rapid. The deviation from the set point is called the dead band.

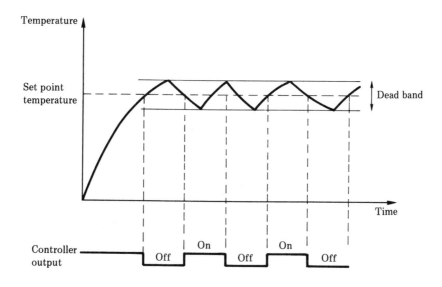

Figure 7.13 Response of an on/off temperature controller.

7.8 Example 7 — Proportional temperature control

7.8.1 THE CONTROL PROBLEM

In this example the PLC is to act as a proportional temperature controller. Figure 7.14 shows how this may be achieved. The system, by using a feedback thermocouple, compares the actual temperature θ_o of the mass being heated with a desired value set by θ_i. An error signal E equal to $(\theta_i - \theta_o)$ is measured. The voltage output to the heater is $K_p E$ where K_p is called the proportional gain constant. As feedback is used the system is described as a closed-loop system.

7.8.2 DESCRIPTION OF CONTROL ELEMENTS

Figure 7.15 shows how the thermocouple and heater are interfaced to the PLC. An ADC is required to read the analogue voltage produced

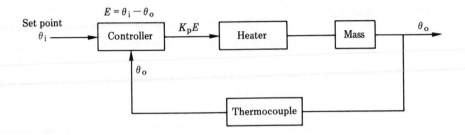

Figure 7.14   Closed-loop temperature control system.

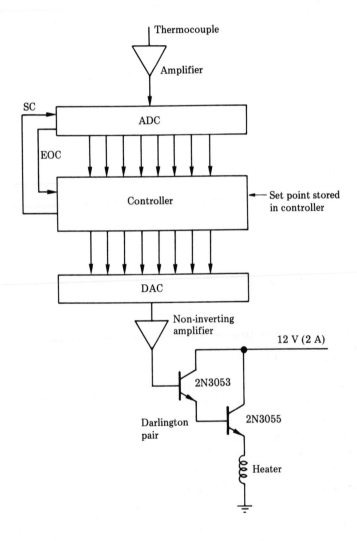

Figure 7.15   Input/output connections for the temperature control system.

from the thermocouple amplifier. It is assumed that the thermocouple amplifier shown in Fig. 4.13 is used. The amplifier's gain is adjusted so that the temperature output matches the ADC input over the required temperature range. The 8-bit ADC then generates a number in the range 0–255 which is assumed to be proportional to temperature over the range required. This number may be scaled to give a value in degrees Kelvin or used directly without scaling.

The heater power supply is controlled by the DAC via a Darlington pair of transistors (the 2N3055 is a power transistor). The 8-bit DAC accepts numbers in the range 0 to 255. Heater power is proportional to the value sent to the DAC's digital input. For example, when zero is sent to the DAC's digital input the heater is off. When 128 is sent to the DAC's digital input the heater is at half power. When 255 is sent to the DAC's digital input the heater is at full power. Using a DAC to vary heater power allows very fine control to be exercised.

7.8.3 PROGRAM SOLUTION

A ladder solution involves a series of functions to perform the following actions: (a) read the ADC, (b) subtract the ADC temperature value from a set point value stored in a data register and (c) multiply the difference value by the proportional gain constant and output to the DAC. The ladder diagram will have the form shown in Fig. 7.16. A flag contact inhibits the DAC if the difference value becomes negative.

A BASIC program for the proportional control of temperature is given below.

```
10 REM Proportional temperature control
11 REM setup ports
12 REM port 1 – handshaking ADC
13 REM port 2 – DAC output
14 REM
20 SETPOINT = 150
30 GAIN = 2
40 T = INP(port 1)
50 ERROR = SETPOINT – T
60 P = GAIN * ERROR
65 IF P < 0 THEN P = 0
70 OUT port 2,P
80 GOTO 40
```

It assumes first that the ADC is connected to a handshaking input port called port 1 and second that the DAC is connected to an output port called port 2. A parallel input port that handshakes data has two special lines called ready and strobe. The ready line is connected to the ADC's start convert pin. The strobe line is connected to the ADC's end

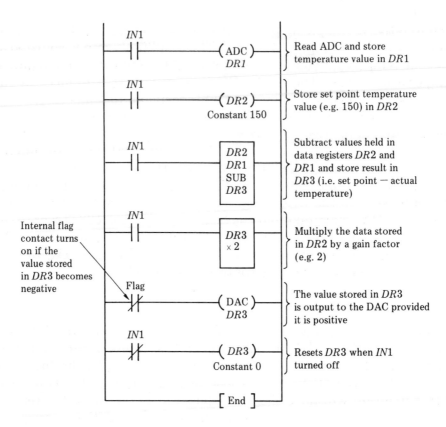

Figure 7.16 Ladder circuit for the proportional control of temperature.

of convert pin. When the port is to be read, a start convert signal is automatically sent to the ADC via the ready line. The port waits for the ADC to generate an end of convert signal on the strobe line before reading the input data.

With proportional control the actual temperature is always offset from the required set point temperature (see Fig. 7.17). To overcome this, add an offset value to line 60 so that it has the form

60 P = GAIN * ERROR + OFFSET.

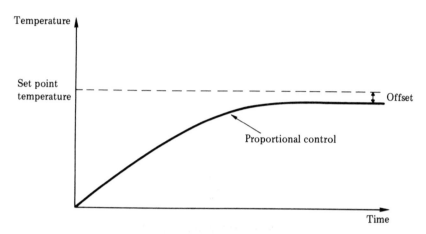

Figure 7.17 Response of a proportional temperature controller.

Questions

1. Develop a ladder program based on a shift register which activates three pneumatic pistons one after the other when a shift command is received.
2. Develop a ladder program which allows a motor to run for two minutes after it has been started.
3. Develop a ladder program which uses a timer to periodically read an analogue signal from an ADC.

8
Case studies

8.1 Introduction

These case studies discuss typical industrial applications of a PLC. They deal with counting and batching components, pick and place operations, a reject system and production line control. A solution in the form of a ladder program is offered for each case study.

8.2 Counting and batching

This study shows how a PLC is used to control a simple machine which counts and batches components moving along a conveyor. A sketch of the batching machine is shown in Fig. 8.1. It is required that ten components be channelled down route A and twenty components down route B. A reset facility is required.

The machine consists of a flap F, which is moved by a piston P, via the pivot linkage L. When solenoid S is energized the piston moves the flap so that components are channelled down route B. When solenoid S is de-energized the piston, being a spring return type, moves the flap so that components are channelled down route A. The microswitch $MS1$ is turned on each time a component passes it. Its function is to count components coming along the conveyor. A proxistor could be used instead of the microswitch.

Figure 8.2 shows how the various components of the batching machine are connected to the input and output ports of the PLC. The start switch is connected to the run input. The microswitch used for counting components is connected to $IN2$. A reset switch is connected to $IN1$. Solenoid S, which controls the flap position, is connected to output port $CR1$.

A ladder program which controls the batching machine is shown in Fig. 8.3. It uses two counters. The first counter is used to direct ten

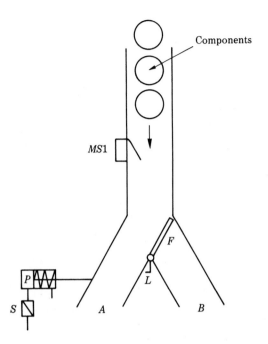

Figure 8.1   Batching machine.

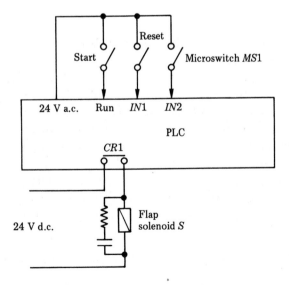

Figure 8.2   Input/output connections for the batching machine.

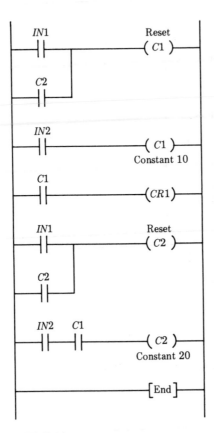

Address	Instruction	Data
0	LOAD	IN1
1	OR	C2
2	RESET	C1
3	LOAD	IN2
4	OUT	C1
5	CONSTANT	10
6	LOAD	C1
7	OUT	CR1
8	LOAD	IN1
9	OR	C2
10	RESET	C2
11	LOAD	IN2
12	AND	C1
13	OUT	C2
14	CONSTANT	20
15	END	

Figure 8.3 Ladder program to control the batching machine.

components down route A. The second counter is used to direct twenty components down route B.

The ladder works as follows. After a microswitch MS1 connected to IN2 has counted ten components the counter C1 closes its contacts and so energizes the flap solenoid via CR1. It also enables counter C2 to start counting components via MS1. After counter C2 has counted twenty components it resets counter C1 and itself. When counter C1 is reset the flap solenoid is de-energized. The scanning ensures that the process is continuously repeated until the run switch is turned off. The reset switch is connected to both of the counter reset lines and enables the user to restart a cycle (i.e. in the case of the flap jamming).

8.3   Pick and place unit

A pick and place unit executes one rotary motion and two linear motions. These motions are defined in Fig. 8.4 and are on/off in nature,

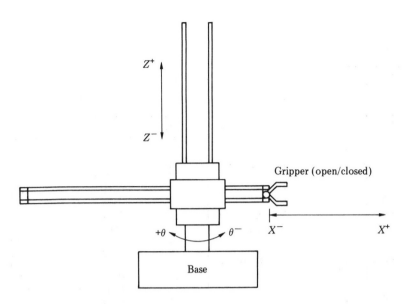

Figure 8.4 Pick and place unit.

being actuated by pneumatic pistons. In addition, a pneumatically operated gripper may be open or shut. As the movements are on/off actions they are often described as bang–bang.

Initially, the gripper is open and positioned at X^- and θ^- position. The PLC is to (a) move the gripper to the X^+ position, (b) close the gripper so that it takes hold of a component, (c) rotate the gripper through 180° to the θ^+ position, (d) release the component and (e) rotate the gripper back to the θ^- position so that the pick and place operation may be repeated.

All movements are controlled by actuating appropriate solenoids. Interfacing involves connecting eight solenoids to eight output relays. It is assumed that the relays are connected so that when energized they cause the following motions.

$CR1$ X^+
$CR2$ X^-
$CR3$ Z^+
$CR4$ Z^-
$CR5$ θ^+
$CR6$ θ^-
$CR7$ close gripper
$CR8$ open gripper

A ladder which performs the control action required is shown in Fig. 8.5. It uses a series of delay-off timer circuits to actuate the solenoids in the required sequence. The timer constants are set to 2 s as this is enough time for the pneumatics to actuate a movement.

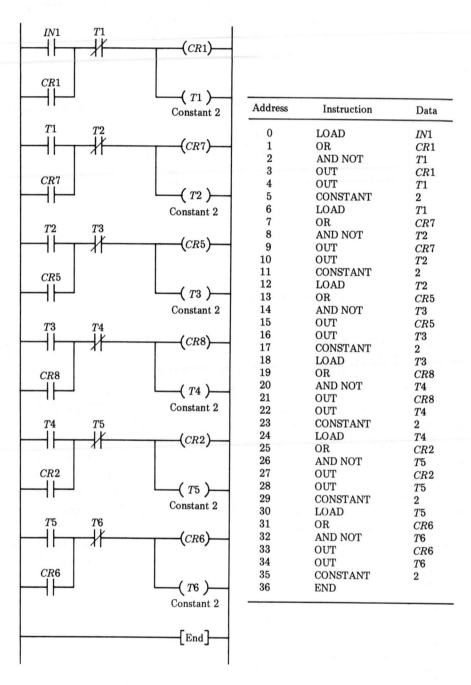

Address	Instruction	Data
0	LOAD	IN1
1	OR	CR1
2	AND NOT	T1
3	OUT	CR1
4	OUT	T1
5	CONSTANT	2
6	LOAD	T1
7	OR	CR7
8	AND NOT	T2
9	OUT	CR7
10	OUT	T2
11	CONSTANT	2
12	LOAD	T2
13	OR	CR5
14	AND NOT	T3
15	OUT	CR5
16	OUT	T3
17	CONSTANT	2
18	LOAD	T3
19	OR	CR8
20	AND NOT	T4
21	OUT	CR8
22	OUT	T4
23	CONSTANT	2
24	LOAD	T4
25	OR	CR2
26	AND NOT	T5
27	OUT	CR2
28	OUT	T5
29	CONSTANT	2
30	LOAD	T5
31	OR	CR6
32	AND NOT	T6
33	OUT	CR6
34	OUT	T6
35	CONSTANT	2
36	END	

Figure 8.5 Ladder program to control the pick and place unit.

It is always good practice to sense whether a movement has taken place. This can be achieved by using limit switches positioned on the unit so that they turn on when a particular movement has occurred. Figure 8.6 shows a ladder diagram which allows two limit switches to control the X^+ and X^- movements.

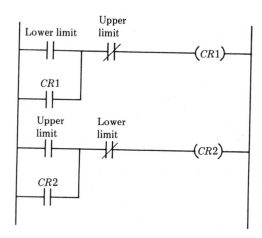

Figure 8.6 Using limit switches to control the X^+ and X^- movement.

8.4 Reject machine

This case study shows how a PLC may be used to detect and reject faulty components. The layout of the system is shown in Fig. 8.7. Components are transported on a conveyor past a retro-reflective type photoelectric switch. The photoelectric switch is positioned at a height (H) above the conveyor, where H represents a tolerance value for component height. Good components pass underneath the photoelectric switch and no signal is generated. Faulty components break the light beam twice as they pass the photoelectric switch.

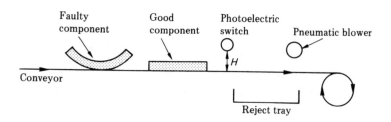

Figure 8.7 Reject system.

When a faulty component is detected a pneumatic blower is to be activated to blow the component into a reject box. All rejected components are to be counted and the current reject count is to be displayed on an external display unit. The operator also requires a reset facility.

The reject control action is controlled by the PLC shown in Fig. 8.8. The normally open contact of the photoelectric switch is connected to $IN1$. A reset switch is connected to $IN2$. The pneumatic blower is actuated by the solenoid connected to $CR1$. The external display unit (a totalizing counter) is connected to $CR2$ and $CR3$. When $CR3$ is energized it clears the display. The unit displays the number of pulses generated by $CR2$. Snubber circuits are essential when interfacing $CR2$ and $CR3$ to the display unit. They prevent false triggering of the display unit caused by contact bounce of the relay contacts.

A ladder program which implements the control action required is shown in Fig. 8.9. A counter $C1$ is used to activate the $CR1$ (i.e. the pneumatic blower) and $CR2$ (i.e. the input to the display unit) when $IN1$ (i.e. the photoelectric switch) generates two pulses. This is because the photoelectric switch counts two pulses when a faulty component is detected. The timer $T1$ keeps the blower energized for 5 s and then resets $C1$. When $IN2$ is momentarily turned on it resets $C1$ and the display unit.

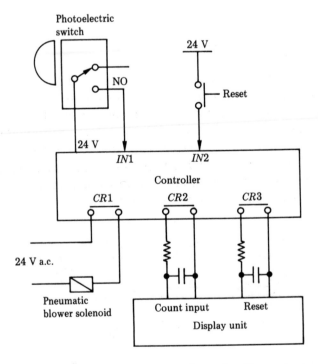

Figure 8.8 Input/output connections for the reject system.

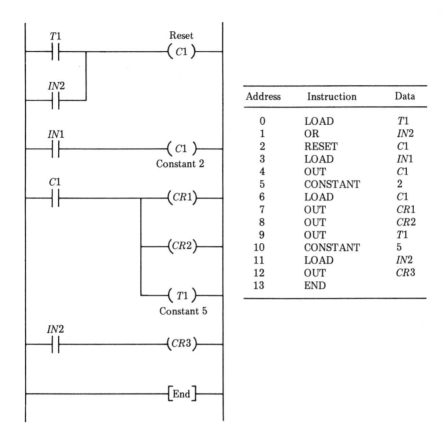

Address	Instruction	Data
0	LOAD	T1
1	OR	IN2
2	RESET	C1
3	LOAD	IN1
4	OUT	C1
5	CONSTANT	2
6	LOAD	C1
7	OUT	CR1
8	OUT	CR2
9	OUT	T1
10	CONSTANT	5
11	LOAD	IN2
12	OUT	CR3
13	END	

Figure 8.9 Ladder program to control the reject system.

8.5   Production line control

This study shows how a PLC may be used to control the production line illustrated in Fig. 8.10. Cans are filled with fluid and capped before passing onto a conveyor. The photoelectric switches $P1$ and $P2$ are used to check that each can has a cap. Photoelectric switch $P3$ provides a trigger for the ink jet printer which prints a batch number on each can. Photoelectric switch $P4$ is used to count three cans into the palletizing machine. The palletizing machine transports three cans through a machine which heat shrinks a plastic wrapping over them. All photoelectric switches on the production line are of the retro-reflective type.

The controller is expected to (a) stop the production line when a can with no cap is detected, (b) activate the ink jet printer when $P3$ detects a can, (c) count three cans using $P4$ before activating the palletizing machine, (d) stop the production line if $P4$ detects a can before palletizing action has finished (the palletizer provides a signal to tell

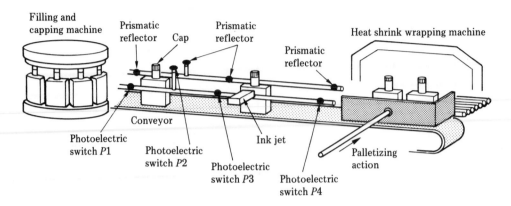

Figure 8.10   Production line control.

the controller it is active) (e) sound an alarm when the production line is stopped and (f) provide an emergency stop pushbutton.

The production line is controlled by the PLC shown in Fig. 8.11. Pushbuttons to start and stop the production line are connected to $IN1$ and $IN2$ respectively. Photoelectric switches $P1$, $P2$, $P3$ and $P4$ are connected to $IN3$, $IN4$, $IN5$ and $IN6$ respectively. The palletizer signal is fed to $IN7$ and turns on when the palletizer is in action. The production line is started via a contactor solenoid connected to $CR1$. The control relays $CR2$, $CR3$ and $CR4$ activate the ink jet, palletizer and operator alarm.

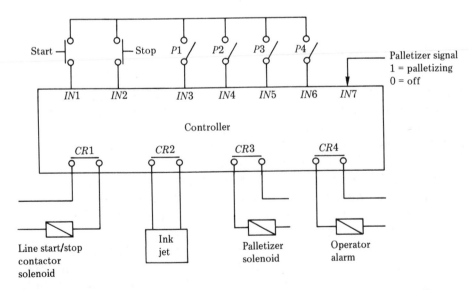

Figure 8.11   Input/output connections for the production line.

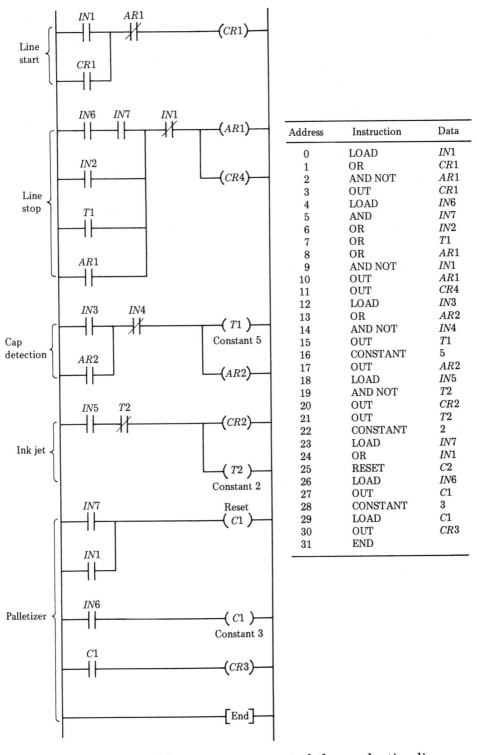

Address	Instruction	Data
0	LOAD	IN1
1	OR	CR1
2	AND NOT	AR1
3	OUT	CR1
4	LOAD	IN6
5	AND	IN7
6	OR	IN2
7	OR	T1
8	OR	AR1
9	AND NOT	IN1
10	OUT	AR1
11	OUT	CR4
12	LOAD	IN3
13	OR	AR2
14	AND NOT	IN4
15	OUT	T1
16	CONSTANT	5
17	OUT	AR2
18	LOAD	IN5
19	AND NOT	T2
20	OUT	CR2
21	OUT	T2
22	CONSTANT	2
23	LOAD	IN7
24	OR	IN1
25	RESET	C2
26	LOAD	IN6
27	OUT	C1
28	CONSTANT	3
29	LOAD	C1
30	OUT	CR3
31	END	

Figure 8.12   Ladder program to control the production line.

A ladder program for controlling the production line is shown in Fig. 8.12. A latch circuit is used to start and stop the production line. The line is stopped by breaking the latch via auxiliary relay $AR1$. The auxiliary relay $AR1$ is energized by the stop pushbutton, the cap detection circuit and when both $P4$ and the palletizer are turned on.

The cap detection circuit makes use of a 5 s timer. Contact $IN3$ ($P1$) sets the timer going. If a cap is not detected in 5 s by $IN4$ ($P2$) the timer's contact stops the line. If a cap is detected by $IN4$ ($P2$) the timer is reset.

The ink jet connected to $CR2$ is energized by $IN5$ ($P3$). Timer $T2$ resets a latch circuit to ensure that the ink jet is energized for 2 s.

The counter $C1$ counts pulses generated by $P4$ connected to $IN6$. When three pulses (three cans) have been counted, the palletizer is activated. The palletizer then turns on $IN7$ which is used to reset the counter. If the palletizer signal is on and a can is detected by $P4$ the line is stopped. This situation occurs when the palletizer becomes jammed.

9
Communications

9.1 Introduction

This chapter is about getting two or more pieces of equipment to transfer information. The information transfer may involve a point to point link such as a computer to PLC or a network of various types of devices. All communication interfaces are either parallel or serial.

9.2 Parallel communications

Parallel communication interfaces use a parallel bus (usually 8-bits wide) to transmit data. They allow data to be transmitted over short distances at high speed. Two common standard parallel communication interfaces are the Centronics and IEEE-488. The Centronics interface is used for connecting printers. The IEEE-488 is mainly used for connecting laboratory instruments to computers.

9.3 Serial communications

A serial interface transmits and receives data one bit at a time. This means that a data byte has to be separated into component bits for transmission and reassembled back again when received. Serial communication interfaces are used for transmitting data over long distances.

9.3.1 RS232

The most common standard serial communications interface is the RS232 which is also called V24 and EIA. The usual RS232 connector is the 25-pin D connector as shown in Fig. 9.1. A minimum cabling configuration will use pins 2, 3 and 7. The connecting lines to pins 2 and 3 normally have to be crossed over so that each device transmits data to a receive pin. A logic 1 is represented by a -12 V and a logic 0 by $+12$ V.

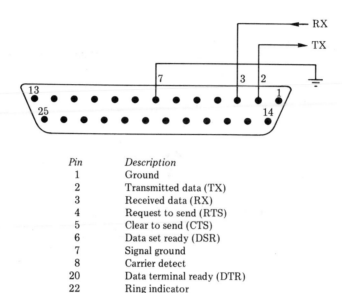

Pin	Description
1	Ground
2	Transmitted data (TX)
3	Received data (RX)
4	Request to send (RTS)
5	Clear to send (CTS)
6	Data set ready (DSR)
7	Signal ground
8	Carrier detect
20	Data terminal ready (DTR)
22	Ring indicator

Figure 9.1   RS232 D connector.

The transmission distance is about 15 m.

With RS232 the user may set several options within the communication process. Both communicating devices must agree on these options, which are as follows.

1. *Baud rate*   This is the operation speed of the serial interface. It is approximately the number of bits transmitted or received per second. Standard baud rates are: 75, 110, 150, 300, 600, 1200, 2400, 4800 and 9600 baud.

2. *Number of bits*   This is the length of the data to be communicated. If data bytes are to be transferred then the number of bits is 8. ASCII codes (these are described below) use 7 bits. Teletype equipment uses 5 bits.

3. *Parity*   Parity is an optional bit added to the data and provides a way of checking whether data has been corrupted. Even parity is when a logic 1 is added to the data to make the number of logic 1s an even number. Odd parity is when a logic 1 is added to the data to make the number of logic 1s an odd number. Space parity is when the parity bit is fixed at logic 0. Mark parity is when the parity bit is fixed at a logic 1.

4. *Stop bits*   Bits added to the end of the data are called stop bits. One or two stop bits may be chosen.

5. *Duplex*   Full duplex requests that a communicating device echo back what it receives. Half duplex tells the communicating device not to echo back what it receives.

6. *Flow control* The simplest way of controlling the flow of data between two pieces of equipment is to set the baud rate to that of the slowest link. Alternatively hardware and software methods may be used. The hardware method requires that the flow control lines 4,5,6,20 of the two RS232 connectors are wired straight through and crossed 4 to 5, 5 to 4, 6 to 20 and 20 to 6. The software methods called XON-XOFF and ETX-ACK use control characters to regulate the flow of data and so do not require pins 4,5,6,20 to be connected.

A typical RS232 signal is illustrated in Fig. 9.2. The data is transmitted at 600 baud. This means that each bit is present for 1.66 ms which equates to 1/600. The first bit is the start bit. The next seven bits represent the data. The least significant bit (LSB) appears first and the most significant bit (MSB) last. The following bit is the parity bit. Even parity is used as a logic 1 is added to the end of the data to make the number of logic 1 s an even number. The last bit is the stop bit.

Data transmitted via an RS232 interface is coded into ASCII codes. ASCII stands for American Standard Code for Information Interchange and is a way of coding characters into a 7-bit form. ASCII codes are listed in Table 9.1. The table shows that the data 100 0011 transmitted in Fig. 9.2 represents the ASCII code for the capital letter C.

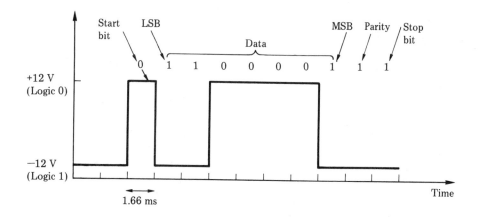

Figure 9.2 RS232 data line at 600 baud. The data format is 7 bits with even parity.

A computer which is used as a programming terminal is normally connected to a PLC via an RS232 link. If programming is done using the computer while the PLC is running and controlling outputs it is termed on-line programming. If the programming is done using the computer with the PLC not controlling outputs it is termed off-line programming.

Table 9.1   ASCII character set (7-bit code)

LS \ MS		0 000	1 001	2 010	3 011	4 100	5 101	6 110	7 111
0	0000	NUL	DLE	SP	0	@	P		p
1	0001	SOH	DC1	!	1	A	Q	à	q
2	0010	STX	DC2	"	2	B	R	b	r
3	0011	ETX	DC3	#	3	C	S	c	s
4	0100	EOT	DC4	$	4	D	T	d	t
5	0101	ENQ	NAK	%	5	E	U	e	u
6	0110	ACK	SYN	&	6	F	V	f	v
7	0111	BEL	ETB	'	7	G	W	g	w
8	1000	BS	CAN	(	8	H	X	h	x
9	1001	HT	EM	)	9	I	Y	i	y
A	1010	LF	SUB	*	:	J	Z	j	z
B	1011	VT	ESC	+	;	K	[	k	{
C	1100	FF	FS	,	<	L	\	l	|
D	1101	CR	GS	−	=	M	]	m	}
E	1110	SO	RS	.	>	N	↑	n	~
F	1111	SI	US	/	?	O	—	o	DEL

CONTROL CODES

NUL	NULL		SI	SHIFT IN
SOH	START OF HEADER		DLE	DATA LINE ESCAPE
STX	START OF TEXT		DC	DEVICE CONTROL
ETX	END OF TEXT		NAK	NEGATIVE ACKNOWLEDGE
EOT	END OF TRANSMISSION		SYN	SYNCHRONOUS IDLE
ENQ	ENQUIRY		ETB	END OF TRANSMISSION BLOCK
ACK	ACKNOWLEDGE		CAN	CANCEL
BEL	BELL		EM	END OF MEDIUM
BS	BACKSPACE		SUB	SUBSTITUTE
HT	HORIZONTAL TAB		ESC	ESCAPE
LF	LINE FEED		FS	FILE SEPARATOR
VT	VERTICAL TAB		GS	GROUP SEPARATOR
FF	FORM FEED		RS	RECORD SEPARATOR
CR	CARRIAGE RETURN		US	UNIT SEPARATOR
SO	SHIFT OUT		SP	SPACE

9.3.2   RS422, RS423 AND RS485

Standards such as RS422 and RS423 are similar to RS232 although voltage levels for the states 1 and 0 differ. An RS485 port is set up in a similar way to an RS232 (i.e. baud rate, data bits, stop bits and parity must be agreed). However, with RS485 the communication is multi-point rather than point to point.

9.4   Local area networks (LANS)

A local area network (LAN) allows a set of PLCs and other devices to be connected together so that they can exchange information. The term local is used because the hardwired link has a limited range. Usually the range is large enough to service a medium-sized factory (500 to 1000 m). Networks which are used for long-distance communications are called wide area networks or WANS.

A network consists of a number of active points (e.g. PLCs) which are called nodes. There are various ways in which nodes can be arranged, depending upon whether the network uses a series of point to point links, a central cable with spurs, or links which make up a ring. Signals transmitted may be any of the following.

1. *Baseband* Baseband systems simply send a digital signal along the connecting cable. Noise immunity can be poor.
2. *Single channel modulated* The signal is superimposed on top of a high-frequency sine wave signal called the carrier. The process of altering a wave so that it carries information is called modulation.
3. *Broadband* Broadband systems modulate a carrier signal so that information is carried in separate frequency bands called channels.

The layout of a typical PLC network is illustrated in Fig. 9.3. It consists of a central network cable with the PLCs connected to it by spurs. The network cable may be fibre optic, coaxial cable or a twisted wire pair. A

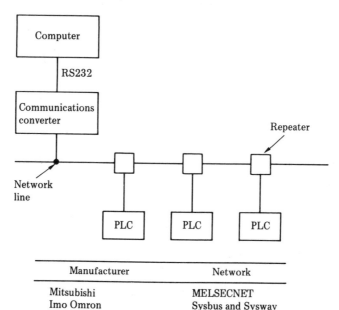

Manufacturer	Network
Mitsubishi	MELSECNET
Imo Omron	Sysbus and Sysway
Toshiba	Tosh-line 30
Allen Bradley	Data Highway
Gould Electronics	Modbus
GEC Industrial Controls	Starnet

Figure 9.3 PLC network.

fibre-optic link is preferred because it is not affected by electromagnetic interference and other types of noise generated in a factory. Repeaters are used throughout the network to boost signal levels.

All networks use a protocol which allows nodes (i.e. PLCs) to communicate without crosstalk. Many PLC manufacturers have developed their own network protocol for their own equipment. These are called proprietary networks. Examples of proprietary networks are listed in Fig. 9.3. Generally the user of a PLC network system does not need to be concerned with the technical details of the network interface or protocol.

In Fig. 9.3, the computer is linked to the network via a communications converter. This converts an RS232 signal into a network signal.

9.4.1 RESPONSE TIME OF NETWORK

The response time of a network (also called the access time) is the time taken for two nodes to communicate. It effectively gives a measure of how fast data can be transferred from one PLC to another. The response time increases as the number of active nodes or PLCs on the network increases. Typically the response time is about 10 ms.

It is essential to know a precise value for the response time if two or more PLCs are to work in unison on a time-critical control application. If the network operates too slowly the control action will fail. Emergency stop signals should never be sent on a network but should be hardwired.

9.4.2 NETWORK STANDARDS

Most PLC networks use a data protocol system developed by the PLC manufacturer. This means that two PLCs from different manufacturers cannot be networked together. Various standards have been suggested which, if adopted by different manufacturers, would overcome this problem. Two standards are now accepted by IEEE (Institute of Electrical and Electronic Engineers). These are as follows.

1. *IEEE 802.3*   This is the Ethernet standard developed by the Xerox Corporation. It is a baseband system which uses single coax for the connecting cable. It uses a protocol called CSMA/CD (Carrier Sense Multiple Access with Collision).

2. *IEEE 802.4*   This standard has been developed by a number of companies and used as a Manufacturing Automation Protocol (MAP). It is a broadband system which uses coaxial cabling. It uses a protocol called token passing. It is a high performance LAN which allows the user to predict the response time of the network.

Questions

1. Convert the message 'on-line' into ASCII codes.
2. What type of link is IEEE-488?
3. What type of link is RS232?
4. What is meant by (a) parity, (b) baud and (c) stop bits?
5. Explain the difference between a broadband and baseband LAN.

Appendix 1
Answers to selected questions

Chapter 1

Q1. NOT
Q2. AND
Q3. OR
Q4. NO = || NC = ⊬
Q5.–7. See Chapter 6

Chapter 2

Q1.(a) Volatility
   (b) Electrically erasable
   (c) Stores the logic states of input/output devices
   (d) Shows how memory is used
Q2. 2K = 2×1024 bytes or 2×1024×8 bits. 2048 instruction codes.
Q3. CPU
Q4.(a) Information to be manipulated by the program
   (b) A register in which bits can be shifted
   (c) A status bit indicating that some condition has occurred
   (d) Memory element
   (e) An error caused by the scan
Q5.(a) 13 octal; B hexadecimal; 11 decimal
   (b) 321 octal; D1 hexadecimal; 209 decimal
   (c) 125252 octal; AAAA hexadecimal; 43690 decimal
Q6. 1100100001
Q7. 1 A; 10 A

Chapter 3

Q3. SSR, contactors and relays
Q5. Noise suppressor

Chapter 4

Q3. 5 V
Q4. Use a multiplexer
Q5. Use a potential divider

Chapter 5

Q4.

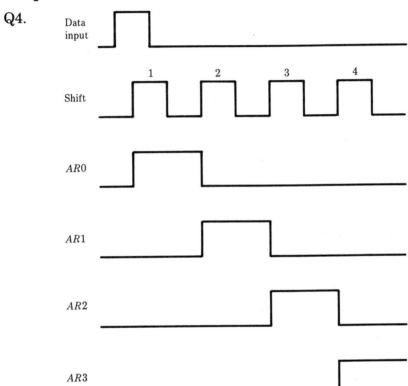

Chapter 6

Q1.

Address	Instruction	Data
0	LOAD	IN1
1	LOAD	IN2
2	AND NOT	IN3
3	OR BLOCK	
4	OUT	CR1

Q2.

Address	Instruction	Data
0	LOAD	IN2
1	AND NOT	IN3
2	OR	IN1
3	OUT	CR1

The circuit of Question 2 is best, as this eliminates the need for an OR BLOCK instruction. This saves scan time.

Q3.

Address	Instruction	Data
0	LOAD	IN1
1	OR NOT	IN2
2	OR	IN3
3	OUT	CR1

Q4.

Address	Instruction	Data
0	LOAD	IN1
1	AND	IN2
2	LOAD	IN3
3	AND	IN4
4	OR	IN5
5	AND BLOCK	
6	OUT	CR1

Q5.

Address	Instruction	Data
0	LOAD	IN1
1	AND	IN2
2	LOAD	IN3
3	AND	IN4
4	LOAD	IN5
5	AND	IN6
6	OR BLOCK	
7	AND BLOCK	
8	OUT	CR1

Q6.

Address	Instruction	Data
0	LOAD	IN1
1	OUT	CR1
2	AND	IN2
3	OUT	CR2
4	AND	IN3
5	OUT	CR3

Chapter 9

Q1. 6f, 6e, 2d, 6c, 69, 6e, 65
Q2. Parallel
Q3. Serial

Appendix 2
Glossary of terms and abbreviations

AC Accumulated count.

a.c. Alternating current.

ADC Analogue to digital converter. A device for converting an analogue signal to an equivalent digital value.

address Numerical designation of a storage location or port.

algorithm A method of computation for solving a problem.

ALU Arithmetic logic unit (part of the CPU).

analogue signal A continuously varying signal.

ASCII American Standard Code for Information Interchange.

auxiliary relay Internal memory element which may be manipulated by the user's program.

back e.m.f. A voltage that opposes a supply voltage.

bang-bang control Simple on/off control.

BASIC Beginner's All Purpose Symbolic Instruction Code. BASIC is a simple to use high-level computer programming language.

baud Speed unit for serial transmission equal to one signal change per second (bits/second).

BCD Binary coded decimal.

bipolar voltage A voltage which may be either positive or negative in sign.

bit An abbreviation for Binary digIT (logic 1 or 0).

block A circuit consisting of two subcircuits.

buffer An interface which stores data on a temporary basis.

bus A set of parallel wires used for carrying information.

byte Eight bits of data.

CAD Computer-aided design.

CAM Computer-aided manufacture.

CIM   Computer-integrated manufacture.

clock   A device which generates periodic signals.

closed-loop system   A system in which the output is fed back and compared with the input.

coil   Control relay or special function.

comparator   A device or ladder function which compares two signals or numbers by subtracting one from the other.

compiler   A program to translate a high-level language into machine code.

contact   An NO or NC input.

counter   A ladder function or hardware device for counting contact operations.

CPU   Central processing unit.

CR   Control relay.

DAC   Digital to analogue converter. A device for converting digital data into an equivalent analogue value.

DATA   Information.

DC   Down counter. A counter which counts from a preset value down to zero.

d.c.   Direct current.

delay-off timer   A timer which, when started, turns on for a fixed period of time (i.e. delays before turning off).

delay-on timer   A timer which, when started, waits for a fixed period of time before turning on (i.e. delays before turning on).

digital signal   A signal which is either on or off.

drum sequencer   A type of counter used for step sequencing.

EEPROM   Electrically erasable programmable ROM.

e.m.f.   Electromotive force. The voltage of an electrical energy source.

enable   A contact (or contacts) which inhibits a function.

EPROM   Electrically programmable ROM.

FET   Field effect transistor.

flag   A bit state (0 or 1) used to indicate that some condition has occurred.

hardware   Physical equipment.

hexadecimal   A number system which uses sixteen digits, i.e. 0 to 9 and A to F (representing 10 to 15).

image memory   Memory which stores an image of the I/O devices.

interface   A circuit which acts as a bridge between two systems.

I/O   Input/output.
ISO   International Standards Organisation.

ladder diagram   A circuit program drawn as a series of contacts, coils and function blocks.
LAN   Local area network. A system of interconnecting PLCs or other devices.
latch   A circuit which keeps an output energized despite any changes in the input. Storage action.
logic   A method of solving problems using AND, OR, XOR and NOT functions.

machine code   Binary numbers which represent CPU instructions.
mask   A pattern of bits which, when logically ANDed with some other value, isolates (clears) those bits not being considered.
message display   An intelligent unit which can store pre-programmed messages.
mnemonic   An abbreviation for an instruction code (e.g. LD for LOAD).
module   Unit or plug-in board which forms part of a system.

NC   (1) Normally closed. (2) Numerical control. A system which uses numerical data to control the movement of a machine tool for the purpose of manufacturing a part.
NO   Normally open.
node   (1) A point on a ladder diagram where two or more connections meet. (2) A network device.

octal   Number system which uses eight numbers, i.e. 0 to 7.
OEM   Original equipment manufacturer.
open-loop system   A system which does not use feedback.
operating system   Program supplied by the manufacturer which governs overall operation.

parallel port   A port which transfers several bits in parallel (usually 8).
PC   Programmable controller (PLC) or personal computer.
photoelectric switch   A light-sensitive device capable of detecting the interruption or completion of a light beam.
PLC   Programmable logic controller. A programmable logic unit used for control.
port   An I/O interface.
preset value   The limiting value for ladder timers and counters.
process control   Analogue control of an industrial process.

program   Sequence of instructions.
protocol   Set of rules used by two or more devices which allow information to be exchanged.

RAM   Random access memory.
real time   Working at the same rate as an ongoing process.
register   Temporary storage location.
relay   Switch controlled by an electromagnet.
response time   The time a system takes to respond to a given event.
retentive or **holding relay**   An auxiliary relay which is battery backed.
ROM   Read only memory.
RS232, RS422, RS423, RS485   Types of standard serial communications ports.

scan   The process of reading image memory and the memory which stores the ladder program instructions.
scan rate   Speed of scanning a given amount of memory (usually 1K).
scan time   The time to scan the whole ladder diagram and update the logic states of the inputs and outputs.
sequence function   Function which executes several steps in sequence.
sink   To take a current from a terminal to earth.
snubber   A snubber is an *RC* circuit. It is connected across a pair of contacts and acts as an electrical noise filter.
software   Programs.
source   To provide a current.
SPDT   Single pole double throw.
SPST   Single pole single throw.
subroutine   A complete program which may be called by another program.

throughput   The number of events a controller can process in a given time.
thyristor   A diode which is triggered into conduction by an additional connection called a gate.
transistor   A device which can be used as an electrically operated switch or as an amplifier.
triac   A device similar to the thyristor but which conducts a.c. when the gate is triggered.
TTL   Transistor–transistor logic.

UC Up counter. A counter which counts up from zero to a preset value.

unipolar voltage A voltage which does not change sign.

VDU Visual display unit.

vision module Camera system with image-processing software stored in ROM.

Appendix 3
Mitsubishi and Omron PLCs

A3.1 Comparison of programming instructions

The table below compares instructions used by Mitsubishi (F series PLCs) and Omron (Sysmac C20K/C28K/C40K PLCs) with the general instructions used in this book.

Instruction[1]	Mitsubishi (F series)	Omron (Sysmac)	Description
LOAD	LD	LD	Load contact
AND	AND	AND	Logical AND
OR	OR	OR	Logical OR
NOT	I	NOT	Inversion
LOAD NOT	LDI	LD NOT	Load inverse
AND NOT	ANI	AND NOT	AND NOT operation
OR NOT	ORI	OR NOT	OR NOT operation
AND BLOCK	ANB	AND LD	AND two subcircuits
OR BLOCK	ORB	OR LD	OR two subcircuits
OUT	OUT	OUT	Output
		TIM[2]	Timer output
		CNT[2]	Counter output
RESET	RST		Reset shift register/counter
SHIFT	SHF		Shift
CONSTANT	K	#	Insert constant
END	END	END	End ladder

1 General instruction used throughout this book.
2 Omron PLCs use the instructions TIM (an abbreviation for timer) and CNT (an abbreviation for counter) for coding timers and counters.

A3.2 Basic programming instructions and data used by Omron Sysmac PLCs C20K/C28K/C40K

Instruction	Symbol	Mnemonic	Data	Data contents
LOAD	⊢ ⊣├─	LD	Point No.	
LOAD NOT	⊢ ⊬─	LD NOT	Point No.	
AND	─⊣├─	AND	Point No.	Point No.
AND NOT	─⊬─	AND NOT	Point No.	0000 to 1907 HR 000 to 915 TIM/CNT 00 to 47 TR 0 to 7 (LD)
OR	⊣├─	OR	Point No.	
OR NOT	⊬─	OR NOT	Point No.	
AND LOAD		AND LD		
OR LOAD		OR LD		
OUT	─○	OUT	Point No.	Point No. 0500 to 1807 HR 000 to 915 TR 0 to 7 (OUT)
OUT NOT	─⊘	OUT NOT	Point No.	
TIMER	─(TIM)	TIM / Set value — Unit:0.1 s Accuracy: +0, −0.1 s	Timer No.	No. TIM/CNT 00 to 47 Set value
COUNTER	CP / R — CNT	CNT / Set value	Counter No.	#0000 to 9999 External setting: 00 to 17 HR CH 0 to 9

DATA POINTS

With Omron PLCs a point number (i.e. data that is substituted for notation such as *IN*1, *T*1, *C*1, etc.) consists of a 4-digit number. The first two digits define a channel consisting of 16 points. The last two digits define a point within a channel. For example, the point number 0101 is interpreted as the second point (00 is the first) within channel 01.

RELAY NUMBERS

Name	CH no. and point no.									
	CH	00	CH	01	CH	02	CH	03	CH	04
	00	08	00	08	00	08	00	08	00	08
	01	09	01	09	01	09	01	09	01	09
	02	10	02	10	02	10	02	10	02	10
	03	11	03	11	03	11	03	11	03	11
	04	12	04	12	04	12	04	12	04	12
	05	13	05	13	05	13	05	13	05	13
	06	14	06	14	06	14	06	14	06	14
Holding relays	07	15	07	15	07	15	07	15	07	15
(HR0000 to 0915)	CH	05	CH	06	CH	07	CH	08	CH	09
	00	08	00	08	00	08	00	08	00	08
	01	09	01	09	01	09	01	09	01	09
	02	10	02	10	02	10	02	10	02	10
	03	11	03	11	03	11	03	11	03	11
	04	12	04	12	04	12	04	12	04	12
	05	13	05	13	05	13	05	13	05	13
	06	14	06	14	06	14	06	14	06	14
	07	15	07	15	07	15	07	15	07	15

Name	CH no. and point no.									
	CH00 (input)		CH01 (output)		CH02 (input)		CH03 (output)		CH04 (input)	
Input/output points (0000 to 0915)	00	08	00	08	00	08	00	08	00	08
	01	09	01	09	01	09	01	09	01	09
	02	10	02	10	02	10	02	10	02	10
	03	11	03	11	03	11	03	11	03	11
	04	12	04	12	04	12	04	12	04	12
	05	13	05	13	05	13	05	13	05	13
	06	14	06	14	06	14	06	14	06	14
	07	15	07	15	07	15	07	15	07	15
	CH05 (output)		CH06 (input)		CH07 (output)		CH08 (input)		CH09 (output)	
	00	08	00	08	00	08	00	08	00	08
	01	09	01	09	01	09	01	09	01	09
	02	10	02	10	02	10	02	10	02	10
	03	11	03	11	03	11	03	11	03	11
	04	12	04	12	04	12	04	12	04	12
	05	13	05	13	05	13	05	13	05	13
	06	14	06	14	06	14	06	14	06	14
	07	15	07	15	07	15	07	15	07	15
	CH 10		CH 11		CH 12		CH 13		CH 14	
	00	08	00	08	00	08	00	08	00	08
	01	09	01	09	01	09	01	09	01	09
	02	10	02	10	02	10	02	10	02	10
	03	11	03	11	03	11	03	11	03	11
	04	12	04	12	04	12	04	12	04	12
	05	13	05	13	05	13	05	13	05	13
	06	14	06	14	06	14	06	14	06	14
	07	15	07	15	07	15	07	15	07	15
	CH 15		CH 16		CH 17		CH 18			
Internal auxiliary relays (1000 to 1807)	00	08	00	08	00	08	00			
	01	09	01	09	01	09	01			
	02	10	02	10	02	10	02			
	03	11	03	11	03	11	03			
	04	12	04	12	04	12	04			
	05	13	05	13	05	13	05			
	06	14	06	14	06	14	06			
	07	15	07	15	07	15	07			

Name	TIM/CNT no.					
Timer/counter (TIM/CNT00 to 47)	00	08	16	24	32	40
	01	09	17	25	33	41
	02	10	18	26	34	42
	03	11	19	27	35	43
	04	12	20	28	36	44
	05	13	21	29	37	45
	06	14	22	30	38	
	07	15	23	31	39	

Name	DM CH no.[2]							
Data memory channel (DM00 to 63)	00	08	16	24	32	40	48	56
	01	09	17	25	33	41	49	57
	02	10	18	26	34	42	50	58
	03	11	19	27	35	43	51	59
	04	12	20	28	36	44	52	60
	05	13	21	29	37	45	53	61
	06	14	22	30	38	46	54	62
	07	15	23	31	39	47		63

[1] TIM/CNT 46 and 47 are reserved for RDM and HDM, respectively.
[2] DM CHs 00–31 are reserved for RDM and DM CHs 32–62 for HDM.

Name	No. of points	Point no.	
Temporary memory relays (TR)	8	TR	0
			1
			2
			3
			4
			5
			6
			7

Note: Points 0000, 0001, and 1804 to 1807 are reserved when HDM and RDM is used.

A3.3 Basic programming instructions used by Mitsubishi F series

INSTRUCTIONS AND EXECUTION TIME

Instructions	Designation	Object factors	Execution time[1]		General functions	
			ON	OFF		
LD	Load	X, Y, M, T, C, S	5.4		Start of logical operation	Normally open contact
LDI	Load Inverse		5.4			Normally close contact
AND	AND		4.2		Logical product (series contact)	Normally open contact
ANI	AND Inverse		4.2			Normally close contact
OR	OR		4.2		Logical sum (parallel contact)	Normally open contact
ORI	OR Inverse		4.2			Normally close contact
ORB	OR Block	None	3.6		Parallel connection of circuit block	
ANB	AND Block				Series connection of circuit block	
OUT	OUT	Y	34.5	34.5	Coil drive instruction	
		M	31.5	31.5		
		S	36.3	48.8		
		T-K	108	142		
		C-K	120	72		
		F671 ~ F675 - K	126	58.9		
PLS	Pulse	M100 ~ M377	49.4	47.0	Rising pulse generating instruction	
SFT	Shift	M100, 120, 140, 160, M200, 220, 240, 260, M300, 320, 340, 360	70.2	50.0	Shift register 1-bit shift instruction	
RST	Reset		63.7	51.8	Reset instruction for shift register, counter	
		C (excluding C661)	44.6	41.7		
S	Set	Y	35.7	29.8	Operation holding coil drive instruction[4]	
		M200 ~ M377	32.7	26.2		
		S	44.6	38.1		
R	Reset	Y	38.1	28.0	Operation holding reset coil drive instruction	
		M200 ~ M377	35.1	25.0		
		S	50.6	32.7		
MC	Master Control	M100 ~ M177	23.8		Common series contact	
MCR	Master Control Reset		3.0		Reset of common series contact	
CJP	Conditional Jump	700 ~ 777	55.4	28.0	Conditional jump	
EJP	End of Jump		0		Designation of conditional jump destination	
NOP	Nop	None	0		None-processing	
END	End		1101[2]		Program end	
STL	Step ladder	S600 ~ S647	14.3 + 69n[3]		Start of step ladder	
RET	Return		14.3		End of step ladder	

[1] It is estimated that one execution cycle time is K times of total execution time calculated from step 0 to END.

$$K = 1.2 + \frac{0.15}{①} + \frac{(0.16}{②} + \frac{0.02}{③} + \frac{0.04)}{④} \frac{f}{⑤}$$

(1) In case T650 T657 are used
(2) In case high-speed counter is used
(3) In case F670 K118 is turned on
(4) In case F670 K121 is turned on
(5) Input frequency of high-speed counter (f = 1 for 1kHz)

[2] Input/output processing time is included.
[3] 'n' shows the number of longitudinal connection (parallel joining) for STL instruction.
[4] 51.2 + 31.5n for ON and 36.9 for OFF in the STL circuit block.
   n . . . Number of longitudinal connection (number of parallel joining) for STL instruction.

DATA POINTS

With Mitsubishi PLCs each data point (i.e. data that is substituted for notation such as $IN1$, $C1$ $T1$ etc.) is assigned a unique number as shown by the list of element numbers below. As the table shows the first contact has the element number 000 or 400. Likewise the first timer has the element number 50.

LIST OF ELEMENT NUMBERS

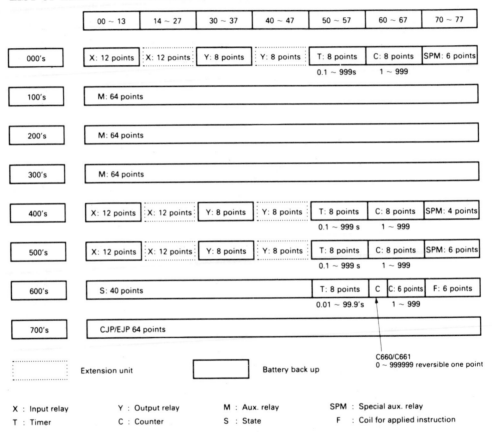

	00 ~ 13	14 ~ 27	30 ~ 37	40 ~ 47	50 ~ 57	60 ~ 67	70 ~ 77
000's	X: 12 points	X: 12 points	Y: 8 points	Y: 8 points	T: 8 points 0.1 ~ 999s	C: 8 points 1 ~ 999	SPM: 6 points
100's	M: 64 points						
200's	M: 64 points						
300's	M: 64 points						
400's	X: 12 points	X: 12 points	Y: 8 points	Y: 8 points	T: 8 points 0.1 ~ 999 s	C: 8 points 1 ~ 999	SPM: 4 points
500's	X: 12 points	X: 12 points	Y: 8 points	Y: 8 points	T: 8 points 0.1 ~ 999 s	C: 8 points 1 ~ 999	SPM: 6 points
600's	S: 40 points				T: 8 points 0.01 ~ 99.9's	C C: 6 points 1 ~ 999	F: 6 points
700's	CJP/EJP 64 points						

C660/C661
0 ~ 999999 reversible one point

⋯⋯ Extension unit   ▢ Battery back up

X : Input relay   Y : Output relay   M : Aux. relay   SPM : Special aux. relay
T : Timer   C : Counter   S : State   F : Coil for applied instruction

Input/output relay Nos. (basic unit)

Basic unit	Input relay Nos.	Output relay Nos.	Extension connector
F₁ – 12M	400 ~ 405 6p	430 ~ 435 6p	400
F₁ – 20M	400 ~ 412 12p	430 ~ 437 8p	400
F₁ – 30M	400 ~ 413 12p 500 ~ 503 4p	430 ~ 437 8p 530 ~ 535 6P	400
F₁ – 40M	400 ~ 413 12p 500 ~ 513 12p	430 ~ 437 8p 530 ~ 537 8p	400 500
F₁ – 60M	000 ~ 013 12p 400 ~ 413 12p 500 ~ 513 12p	030 ~ 037 8p 430, ~ 437 8p 530 ~ 537 8p	000 400 500

p: points

The value in ▢ of extension unit will be "0", "4" or "5", depending upon extension connector Nos. 000, 400 or 500.

Input/output relay Nos. (extension unit)

Extension unit	Input relay Nos.	Output relay Nos.
F – 4T	▢ 20 ~ ▢ 23 4p	▢ 40 ~ ▢ 43 4p
F₂ – 8EY	—	▢ 40 ~ ▢ 47 8p
F₁ – 10E F – 10E	▢ 14 ~ ▢ 17 4p	▢ 40 ~ ▢ 45 6p
F₂ – 12EX	▢ 14 ~ ▢ 27 12p	—
F₁ – 20E F₂ – 20E F – 20E	▢ 14 ~ ▢ 27 12p	▢ 40 ~ ▢ 47 8p
F₁ – 40E F₂ – 40E F – 40E	414 ~ 427 12p 514 ~ 527 12p	440 ~ 447 8p 540 ~ 547 8p
F₁ – 60E F₂ – 60E	014 ~ 027 12p 414 ~ 427 12p 514 ~ 527 12p	040 ~ 047 8p 440 ~ 447 8p 540 ~ 547 8p

p: points

A3.4 Equivalent coding

Some examples of how the general instructions used in this book are translated into equivalent instructions used by Mitsubishi and Omron are shown below.

(a) INPUT/OUTPUT

			Mitsubishi code			Omron code		
0	LOAD	*IN*1	0	LD	400	0	LD	0000
1	OUT	*CR*1	1	OUT	430	1	OUT	0001

(b) COUNTERS

			Mitsubishi code			Omron code		
0	LOAD	*IN*1	0	LD	400	0	LD	0000
1	RESET	*C*1	1	RST	60	1	LD	0100
2	LOAD	*IN*2	2	LD	401	2	CNT	01
3	OUT	*C*1	3	OUT	60	3		#0005
4	CONSTANT	5	4	K	5			

(c) TIMERS

			Mitsubishi code			Omron code		
0	LOAD	*IN*1	0	LD	400	0	LD	0000
1	OUT	*T*1	1	OUT	50	1	TIM	00
2	CONSTANT	5	2	K	5	2		#0005

(d) SHIFT REGISTERS

			Mitsubishi code[1]			Omron code[2]		
0	LOAD	*IN*1	0	LD	400	0	LD	0000
1	OUT	*BIT*0	1	OUT	130	1	LD	0001
2	LOAD	*IN*2	2	LD	401	2	LD	0002
3	SHIFT	*BIT*0	3	SHF	130	3	SHF	01
4	LOAD	*IN*3	4	LD	402	4		01
5	RESET	*BIT*0	5	RST	130			

[1] If the shift register is 8 bits long bit elements range from 130 to 138.
[2] Omron code assumes that 0000 is the data input, 0001 is the shift (clock) input and 0002 is the reset input of the shift register. A start and end channel must be specified. In this example the start channel is 01 and the end channel is 01. As all shift registers are 16 bits long, bit elements range from 0100 to 0115.

Further reading

B. R. Bannister and D. G. Whitehead, *Transducers and Interfacing, Principles and Techniques*. Van Nostrand Reinhold, 1986.

D. V. Hall, *Microprocessors and Interfacing 'Programming and Hardware'*. McGraw-Hill, 1986.

B. E. Jones, *Instrumentation, Measurement and Feedback*. McGraw-Hill, 1977.

M. A. Needler and D. E. Baker, *Digital and Analog Controls*. Prentice Hall, 1985.

S. R. Ruocco, *Robot Sensors and Transducers*. Open University Press, 1987.

Magazines such as *Automation (The Journal of Automated Production), Control Systems, Robotics World, Electrical Review,* and *Professional Engineering (The Magazine of the Institution of Mechanical Engineers)* provide useful reviews of hardware and software.

Index